Physical Anti-Collision in RFID Systems

Xiaolei Yu · Zhimin Zhao · Xuezhou Zhang

Physical Anti-Collision in RFID Systems

Theory and Practice

Xiaolei Yu
Nanjing University of Aeronautics
and Astronautics
Nanjing, Jiangsu, China

Zhimin Zhao
Nanjing University of Aeronautics
and Astronautics
Nanjing, Jiangsu, China

Xuezhou Zhang
Jiangsu Institute of Quality and Standards
Nanjing, Jiangsu, China

ISBN 978-981-16-0837-7 ISBN 978-981-16-0835-3 (eBook)
https://doi.org/10.1007/978-981-16-0835-3

Jointly published with Science Press
The print edition is not for sale in China (Mainland). Customers from China (Mainland) please order the print book from: Science Press.

This Springer imprint is published by the registered company Springer Nature Singapore Pte Ltd.
The registered company address is: 152 Beach Road, #21-01/04 Gateway East, Singapore 189721, Singapore

Preface

The Internet of Things (IoT) is a hot concept and developed rapidly in recent years. It is an important component of the future generation of information systems. The emergence of IoT is another information technology revolution after computer, internet, and mobile communication. As one of the core technologies in the field of IoT perception, Radio Frequency Identification (RFID) is a non-contact automatic identification technology that developed rapidly in the 1990s. It is a non-contact two-way communication of inductive or electromagnetic radiation, using radio waves and microwaves, to achieve purposes of automatic identification of target objects, access to relevant data, and data exchange.

The concept of "physical anti-collision" proposed in this book is relative to the software anti-collision. It is found that the reasons for multi-tag collisions that occur in the frequency bands above Ultra High Frequency (UHF) have both the inherent design of algorithm flaws and physical interferences such as some tags that cannot be identified due to various external physical interferences. Physical anti-collision mainly solves the problem of the low batch recognition success rate caused by the latter. Therefore, physical collision avoidance is defined as the use of physical means to solve the problem of the multi-tag collision caused by non-software factors. The front-end of the physical anti-collision system uses physical means to collect data, for example, image sensors are used to collect the geometric characteristics of multi-label distribution, and the back-end uses neural network algorithms to learn, train, and predict the physical optimal distribution structure, and then adjust the label arrangement through physical means, angle, to achieve the overall optimal recognition performance of the tag group.

This book is based on the basic principles of physics (including electromagnetics, optics, thermodynamics, etc.), with engineering mathematical methods as the core of detection and control algorithm design. Furthermore, this book innovatively applies semi-physical verification and detection technology to the dynamic performance testing of Radio-Frequency Identification (RFID) systems. This book proposes a series of new theories and methods for physical collision prevention, as well as related test verification methods for building semi-physical hardware platforms based on photoelectric sensing technology, which will provide important theories and technical support for the applications of IoT systems in smart logistics, car networking, food traceability, anti-counterfeiting, and other livelihood fields.

This book has the advantages of reasonable chapter arrangement, clear structure, rich content, good illustrations, and texts. It contains not only basic theories about photoelectric sensing, but also related research and analysis of dynamic detection system design and testing, especially "physical anti-collision" is innovative and has application value, which is very suitable for scientific research personnel in the IoT and optoelectronic engineering related industries. After years of hard work, the authors' scientific research team has published more than 30 high-level SCI-indexed papers in related fields, applied for more than 10 authorized invention patents, and obtained important academic achievements including the 16th China Patent Award and Jiangsu Science and Technology Progress Award, four national standards and Jiangsu provincial standards were formulated, which contributed a solid theoretical and practical foundation for the publication of this book.

This book is divided into six chapters. They reflect a frontier innovation research direction of the current research field of IoT applications and RFID system detection, and the latest achievements of the authors' research team in the field of RFID physical anti-collision technology and semi-physical simulation of IoT system. The main contents of this book include an overview of RFID system anti-collision technology, design for RFID system physical anti-collision experimental verification system, physical theory of RFID system physical anti-collision, image theory of RFID system physical anti-collision, optimization algorithm and physical anti-collision of RFID system as well as deep learning and RFID system physical anti-collision.

Nanjing, China
October 2020

Xiaolei Yu
Zhimin Zhao
Xuezhou Zhang

Acknowledgments

This book is a summary of our research results achieved in recent years. Many people contributed to this book in various ways. The authors are indebted to research students, Ph.D. Lin Li, Ph.D. Xiao Zhuang, Ph.D. Zhenlu Liu, M. S. Ke Zhang, M. S. Cheng Ding, M. S. Shanhao Zhou, M. S. Zhuo Zhu at Nanjing University of Aeronautics and Astronautics, for their assistance in pointing out new ideas, preparing the materials, and checking the whole book. Especially, we thank IEEE Fellow & Australian Future Generation Professor-Prof. Jonathan Mantan from the University of Melbourne, Cheung Kong Scholar-Prof. Rushan Chen from Nanjing University of Science and Technology, Prof. Xiang Li and Senior Engineer Yu Huang from Chinese National Quality Supervision and Testing Center for RFID Product (Jiangsu), and all the contributors from around the world who have contributed ideas and materials to this book.

This book is supported by the National Natural Science Foundation of China (Grant No. 61771240), Six Talent Peaks Project in Jiangsu Province (Grant No. XYDXX-058), China Postdoctoral Science Foundation (Grant No. 2015M580422 & 2016T90452), Science and Technology Project of AQSIQ (Grant No. 2017QK117 & 2013QK194) as well as the Fund Project of Jiangsu Engineering Laboratory for Lake Environment Remote Sensing Technologies (Grant No. JSLERS-2018-003). The authors really appreciate the supports.

Contents

Chapter 1
Overview of RFID System Anti-Collision Technology

This book is based on the basic principles of physics (including electromagnetics, optics, thermodynamics, etc.), with engineering mathematical methods as the core of detection and control algorithm design. Furthermore, this book innovatively applies semi-physical verification and detection technology to the dynamic performance testing of radio-frequency identification (RFID) systems. This book proposes a series of new theories and methods for physical collision prevention, as well as related test verification methods for building semi-physical hardware platforms based on photoelectric sensing technology, which will provide important theories and technical supports for the applications of the internet of things (IoT) systems in smart logistics, car networking, food traceability, anti-counterfeiting, and other livelihood fields. The main contributions of this book are as follows:

(1) It introduces a new theory of physical anti-collision for improving the integral performance of multi-tag RFID systems.
(2) It studies the principles and optimization algorithms of innovative multi-tag network topologies, which enhance the sensing capability of the RFID system.
(3) It provides valuable guidance for RFID system physical anti-collision based on deep learning.
(4) It combines physical and electrical expertise from an interdisciplinary standpoint toward the analysis and design of dynamic performance testing systems.

1.1 RFID Collision Problem and Research Progress of Anti-Collision

Internet of things (IoT) as a new technology has developed rapidly in recent years, which is an important part of a new generation of information systems. The emergence of IoT is another information technology revolution after computers, the Internet, and mobile communications [1]. As one of the core technologies in the field of perception of IoT, radio-frequency identification (RFID) technology is a

X. Yu et al., *Physical Anti-Collision in RFID Systems*,
https://doi.org/10.1007/978-981-16-0835-3_1

non-contact automatic identification technology that emerged and developed rapidly in the 1990s. It uses radio waves by induction or electromagnetic wave radiation. The non-contact two-way communication achieves the purpose of automatically identifying target objects, obtaining relevant data, and data exchange. Internationally, RFID systems have been widely used in logistics supply chains, transportation, electronic tickets, production and assembly, anti-counterfeiting and anti-theft, smart agriculture, environmental protection, military defense, and many other fields [2–5].

Major developed countries and regions in the world have successively carried out research on RFID and IoT technology. The National Development and Reform Commission of China included RFID in the first batch of national key industry adjustment and revitalization plans in the electronic information field. In recent years, China has successively carried out the research and development and industrialization of RFID-related technologies, and has initially established a certain foundation in RFID and IoT technology research and development, standard development, industry cultivation, and industry applications. At present, although the annual scale of China's RFID technology market exceeds RMB 10 billion, it ranks third in the world, after the United Kingdom and the United States. However, compared with existing foreign technologies, China's RFID technology foundation is relatively weak compared with developed countries. There is still a gap, especially in the research of chips, tags, and sensor networks compared with developed countries. After years of development, the accuracy and performance stability of low-frequency and high-frequency RFID-related products in China has reached the international level, in UHF and microwave frequency bands, many studies are still in the preliminary explorations and laboratory research stages. Moreover, there is a big gap between indicators such as the sensitivity and reliability of the performance and foreign products. The immaturity of these technologies limits the pace of RFID development, and has become a hot spot and difficulty for breakthroughs in the research of RFID technology in the frequency bands above UHF. With the continuous rapid development of China's economy and the continuous upgrading of the industrial structure, in the large-scale intelligent and information high-tech industry, there are several key technologies problems of RFID systems above the UHF frequency band that urgently need to be solved in the field of RFID application promotion. The obvious asymmetry phenomenon of advanced technology and weak basic research has severely restricted the development of RFID and IoT engineering disciplines.

Since the data transmission of the RFID system is a weak signal wireless communication, the integrity and accuracy of the data will be interfered with by the external environment. The actual communication environment (multipath interference, noise environment, metal structure, liquid environment, etc.) The signal performance has a great impact [6, 7]. With the increase of the working frequency of the RFID system, the batch reading success rate of RFID multi-tags has become the key to the improvement of the system's automatic identification efficiency. In many practical application scenarios of RFID, multiple tags of the same specification may be closely fitted and overlapped. For example, batch initialization read and write operations are carried out after stacking product tags are shipped. In file and book management applications, there are many tags on the surface of the book and these tags may overlap closely. When multiple tags overlap closely, the resonant frequency shift caused by tag coupling and the loss of coupling energy makes the tag unable to read and write normally. This phenomenon is more obvious under the interference of transient electromagnetic waves, which seriously affects the working efficiency of the RFID system. Therefore, the read–write barrier of overlapping tags is a technical bottleneck in RFID industrial applications. Studies have shown that when multiple tags are distributed according to a certain geometric shape, different conditions such as electromagnetic fields, media, and RFID tags' speed, angle, and distance relative to reader antenna in different spaces are all key factors affecting the communication performance between readers and tags [8–11]. In some areas involving public safety, electromagnetic interference with RFID systems has been listed as a "serious and dangerous source of interference". According to a recent study published in The Journal of the American Medical Association (JAMA), electromagnetic interference will seriously damage the RFID-centered medical Internet of Things system in the in-depth treatment test, which was conducted by the Academic Medical Center of the University of Amsterdam in the Netherlands [12]. Pacemakers, defibrillators, dialysis machines, infusion/syringe pumps, and ventilator fans are all medical devices that are susceptible to interference. The interference of electromagnetic waves, especially transient electromagnetic waves, on RFID systems will have a direct physical impact on patients, such as patients. The IV pump or pacemaker in use is stopped and the device issues an incorrect alarm and so on. Therefore, in the intensive care unit and other similar medical environments, the implementation of RFID technology must carry out on-site electromagnetic interference tests in accordance with the latest international standards. Researchers at the RFID laboratory of the University of Wisconsin-Madison have been studying and evaluating the suitability of using RFID technology to track blood products and expensive medical supplies. The Chinese government also attaches great importance to the intelligence of medical production, manufacturing, and transportation. Premier Keqiang Li proposed at the executive meeting of the State Council held in February 2016 that "promoting the intelligent transformation of pharmaceutical production as an important measure for the innovation and upgrading of the pharmaceutical industry" was proposed. In the process of intelligent transformation, safety and reliability are the keys to the application of RFID technology [13, 14]. In addition, in the field of mass production and logistics, in order to improve the degree of automation, many UHF RFID tags are used. When there are multiple RFID tags in the effective range of a reader, because all tags use the same working frequency, the communication between the reader and the tag shares the wireless channel, and sending data at the same time may cause channel contention, data conflict, and transmission issues such as signal interference, leading to information loss, so that the reader cannot correctly identify all tags, this is the collision problem of the RFID multi-tag system. At present, it is the most concerned and the hottest research in the field of RFID technology research and application at home and abroad [15, 16].

There are many factors that cause the collision phenomenon, but electromagnetic interference, static electricity, and environmental factors (such as metals, liquids) are the main reasons. A study found that when the goods with RFID tags pass through the gate of the RFID reader antenna in batches, transient electromagnetic interference is one of the most critical factors affecting the reading performance of the RFID multi-tag system. The collision of the RFID multi-tag system caused by it will directly reduce the working efficiency of the system [17–19]. The impact of transient electromagnetic interference generated by temperature changes in the application environment, various noises, and object media (such as metal) attached to the RFID tag on the tag reading rate and the resulting RFID multi-tag collision. These technical problems have seriously affected the development and application of RFID technology [20].

In summary, modeling of RFID multi-tag collision mechanism based on transient electromagnetic waves and RFID anti-collision dynamic detection technology has become important technical guarantees in the process of RFID technology development and application implementation. Therefore, studying the mechanism of RFID multi-tag collision caused by transient electromagnetic waves and adopting effective anti-collision methods and measures are key issues that have important theoretical value and application prospects and urgently need to be resolved in the promotion and application of RFID multi-tag systems.

This book has conducted in-depth investigations on relevant key technical issues in the early stage, and conducted a lot of preliminary exploratory research work under this background. Through summary and analysis, it is found that the current mechanism of RFID system collision is caused by external interference and anti-collision measures. The main problems in the research are as follows:

(1) At present, the existing research at home and abroad is to carry out theoretical research on RFID multi-tag collision caused by metal, liquid, electrostatic, electromagnetic, and other interference forms from different perspectives. However, a mathematical model and experimental evaluation method that can more systematically and accurately reflect RFID multi-tag collisions caused by interference in practical application scenarios have not been established.
(2) The anti-collision research carried out so far mainly focuses on various anti-collision algorithms at the communication protocol layer (i. e., software anti-collision). But the main idea of these algorithms is to solve the problem of mutual interference between tags, especially transient electromagnetic waves. The harm caused by electromagnetic interference cannot be solved. And the diversity of actual application scenarios makes it impossible to completely solve the collision problem of the RFID system by only using software algorithms.
(3) Once a multi-tag collision occurs, there are currently no effective remedial measures, which greatly reduces the efficiency of automatic identification, especially the impact on the efficiency of logistics transportation in the high-speed dynamic assembly line environment.
(4) Although there are multi-tag anti-collision research and multi-reader anti-collision research, there is no research report on the use of RFID multi-tag-multi (reader) antennas to form a sensor network for collaborative anti-collision.

These problems are the technical keys to be solved urgently, and the researches about them have attracted widespread attention at home and abroad. This book proposes to apply the control and coordination mechanisms (such as neural control, autonomous learning, and self-organization) existing in biological systems to RFID multi-tag and multi-antenna systems in complex interference environments, and to build a bionic sensor network with physical anti-collision capabilities. It intends to solve the technical problems of semi-physical modeling of RFID multi-tag-multi-antenna system collision caused by transient electromagnetic wave interference and the technical problems of RFID multi-tag-multi-antenna sensor network adaptive and cooperative anti-collision.

The autonomous control of the system mainly depends on the active perception and information fusion capabilities of the system itself. It is precisely based on this consideration that in recent years, research on information fusion based on advanced sensors has attracted widespread attention and achieved many research results. However, the current research are no longer satisfied with the research on multi-sensor (or sensor network) capabilities. More attention is paid to the more advanced autonomous information acquisition and fusion capabilities of the system itself, that is, the traditional multi-sensor fusion integration research is further extended to the deeper "Multi-sensory Integration" field.

Various organisms have a wealth of sensory organs, which can sense external stimuli through various senses such as sight, hearing, smell, taste, touch, etc., in order to fully obtain the most valuable environmental information, and the brain and central nervous system Respond effectively (feedback) to the acquired external incentive information. The research of bionic perceptual information fusion theory under changing environment involves many fields such as artificial intelligence, electronic information, cognitive psychology, neurophysiology, etc. It is an important frontier research content in the emerging interdisciplinary-bionic science. The research starts from the basic principles of autonomous perception of organisms, explores the environmental cognitive process of the system, and simulates the neural activities and intelligent behaviors under the active perception mechanism of humans or other organisms. The research focuses on the perception, screening, and understanding of information in the natural environment. The research improves the system's understanding of the environment from a one-sided, discrete, and passive perception level to a global, related, and active perception level. This research explores the establishment of a method system suitable for autonomous perception and control in a dynamic environment, which is a robust modeling requirement for the system to achieve adaptive autonomous control [21–24].

In recent years, foreign researchers have carried out research on machine perception behavior from the perspective of bionics. Many scholars pay attention to the application of artificial neural network methods to the control of robots, and have designed various artificial neural networks to simulate the thinking activities of robots, opening up the relevant research fields of robot bionic perception [25–28]. However, the artificial neural network method does not consider the complex thinking mechanism of the actual biological neural network. Artificial neural networks need training and memory storage. Robots require a lot of time to learn and train, which is not conducive to real-time control; in addition, in many unknown dynamic environments, this type of robot cannot adapt to the constantly changing environment and complete autonomous control. Especially the robot's autonomous orientation problem cannot be solved by the general neural network method. Other theories such as fuzzy control and genetic algorithm have also been introduced into the design of robot behavior controllers or the study of behavior coordination and fusion strategies, but in practical applications, there are generally problems such as poor reliability and weak adaptive ability [29–32].

In 1998, the American scholar Arkin first proposed perception is not an independent process, but a unified whole that works with dynamic systems and control systems [33]. Later researchers also generally believed that perception is an inseparable process from behavior, that is, the behavior needs to provide specific content for the perception process, and perception provides the information needed for motion control. In motion control, the perception ability of the robot can be improved by placing multiple sensors in the appropriate position, which is conducive to the robot selecting important information for specific tasks and discarding less important information. Traditional cognitive science divides the actual world into different categories (perception types) as internal expressions (models) formed by facing the external environment. Researchers of machine perception got inspiration from the perception mechanism of this organism and began to study a new robot perception scheme. Among them, the Swiss Federal Institute of Technology Verschoor and his collaborators proposed a new perception structure called "Distributed Adaptive Control" (DAC) and applied it to the robot in the neural model [34, 35]. The DAC structure is composed of three closely related control layers: Reactive Layer, Adaptive Layer, and Contextual Layer. The reaction layer implements a series of conditioned reflex behaviors based on low-level perception and unconditional input. The adaptive layer associates the system with more complex external stimuli. The background layer establishes a high-level expression through the memory storage structure.

Almost all adaptive behaviors require the fusion and processing of multi-sensory information, and the transformation of this information into a series of purpose-guided behaviors. Scientists have discovered in the study of certain animals in nature that the entire perception process of animals is completed by external (environmental) stimuli and internal (animal body) feedback [36]. The cerebral cortex processes information from recognized targets in the environment. This information is obtained through long-term stimulation (training) of the neural unit receiver. In this process, the overall dynamic behavior constitutes a neural response to external excitation, and then a nonlinear dynamic connection is formed in the cerebral cortex. Freeman, a famous biologist at the University of California, Berkeley, and his collaborators have discovered the dynamics of animal cerebral cortex in the process of perceptual processing through long-term experimental research, and put forward the perceptual dynamics theory, which considers brain activity. It can be expressed by the behavior of Chaotic Dynamics [37–39]. They did many interesting experiments with rabbits, such as letting rabbits inhale various odors, and then observe them through EEG scanners. They evaluated the electrical stimulation response of the olfactory bulb in the rabbits' brains. The electric waves of the olfactory bulb exhibit complex dynamics. They concluded that the internal neural expression pattern produced by an external stimulus is the result of complex dynamic behaviors in the brain (perception) cortex. Freeman further proposed that the dynamic behavior of the olfactory bulb in the brain can be described by a three-dimensional chaotic attractor with multiple volumes, and these "volume" structures can be considered as memory trajectories formed through long-term autonomous learning by the brain's nerve tissue [40]. In the absence of external stimuli, the system is in a high-dimensional iterative search mode, and the search trajectories are different volumes. But once the excitation signal is received, the dynamic behavior of the system appears to be restricted to a certain "volume" for periodic vibration, and this specific volume just reflects the characteristics of the external excitation signal. If the excitation signal input to the system stops, the system will immediately switch to the high-dimensional iterative search mode. According to their work, the activity of the nervous system is maintained in a chaotic state until the sensor interrupts this behavior. The result of this process is that a strange attractor representing the newly introduced excitation appears, and the function of chaos is to provide the system with sufficient flexibility and robustness during the migration process of different perception states. Based on analyzing and summarizing a large amount of experimental data, Freeman proposed a dynamic model of the animal olfactory system, called "K-sets" [37]. The scholars Harter and Kozma of the University of Memphis in the United States gave the discrete form of the K-sets model in 2005 and applied it in the field of navigation control [41]. The selection of control parameters in the system is obtained based on evolutionary learning, and at the same time, an unsupervised learning strategy is adopted. Subsequently, Fukui University Islam, Italy University of Catania Arena, and American scholar Kozma, etc. also researched on navigation robots based on the K-sets model from different angles and different fields [42–44]. It is worth noting that due to the academic frontier and interdisciplinary nature of the research in the above fields, related research papers have been published in the top international journals such as Nature and Science in recent years, which have attracted widespread attention from scholars in many fields [34, 45–47]. The EU framework program SPARK, NASA's Mars Exploration Program, and the Australian Collaborative Research Strategic Project CRC, etc. all take "intelligent information perception and fusion" related research as a key support research direction, and apply it to spacecraft, robots, and intelligence. In transportation and other related fields, other major developed countries have successively carried out exploratory research work in this field. "Intelligent Perception Technology" and "Intelligent Information Processing" are also listed as frontier research fields and priority development themes in the outline of my country's medium and long-term scientific and technological development plan (2006–2020). However, the problem of the system's weak adaptability to the environment has not been well resolved, which has bothered and affected the substantial progress of the technology. During his Ph. D. study in Australia, the first author of this book, under the guidance and cooperation of Prof. Jonathan Mantan from the Department of Electronic Engineering of the University of Melbourne and Dr. Robert Stewart from the Australian Commonwealth Scientific and Industrial Research Organization

(CSIRO), focused on photoelectric sensing and bionic control of moving bodies, did some preliminary explorations, and obtained some meaningful research results [48–53]. After returning to China to build a research team, the team members continued to carry out more in-depth research in related fields. Years of research accumulation laid an important foundation for the research contributions [54–62].

Autonomous perception is an important adaptive ability that exists in the process of biological evolution. The research results of the formation mechanism of the biological body's cognitive response to the environment are applied to the autonomous control of RFID multi-tag-multi-antenna systems. It studies the autonomous sensing of RFID multi-tag-multi-antenna systems and the active control of external stimuli in unknown environments. It stimulates the unconscious reaction mechanism and conscious perception behavior of living organisms in the process of interacting with the environment, and verifies a series of concepts, phenomena, and mechanisms observed and proposed in neuroscience experiments through a bionic intelligent system. These will help establish a robust RFID multi-tag-multi-antenna system autonomous sensing and control system in a complex interference environment.

Based on the current research status and development trends at home and abroad, in response to the problem of poor anti-collision performance and weak adaptive anti-interference ability of RFID multi-tag and multi-antenna systems under transient electromagnetic interference, this book creatively proposes: draw lessons from basic research in biology and neuroscience. Achievements and related theories of the multi-volume chaotic self-organization control mechanism regard the RFID multi-tag-multi-antenna wireless sensor system as a living organism, and regard self-sensing and self-organization as the RFID multi-tag-multi-antenna wireless sensor the vital characteristics of the system. This book uses the strongly nonlinear dynamic behavior- "chaos" to simulate the "perception" activity of the system, through the self-organization adjustment of the relative position and angle of the multi-tag-multi-antenna, and explores how to integrate the loop of the perception-behavior dynamic of the organism. The self-organizing regulation mechanism and law are applied to the physical collision avoidance research of RFID multi-tag-multi-antenna wireless sensor system. This book solves the bottleneck problem of RFID multi-tag collision caused by transient electromagnetic waves in the practical application of RFID multi-tag-multi- antenna system, which is also the fundamental source of the academic thoughts of this book. The research in this book will play an important role in promoting the development of a new generation of multi-sensor fusion information systems, and will have important scientific significance and broad application prospects for solving the bottleneck problem of the development of RFID and Internet of Things engineering technology.

1.2 RFID System Anti-Collision Technology

Compared with other identification technologies, one of the important advantages of RFID technology is that it can realize multi-objective simultaneous recognition. In most RFID identification scenarios, there are two common scenarios. One is that there are multiple RFID tags in the reading range of a reader, which is called tag collision. Another situation is that a tag may be read by multiple readers at the same time, meaning that the scope of the readers overlaps.

In order to realize simultaneous recognition of multiple objects, it is necessary to solve the problem of signal interference when multiple tags correspond to one or more readers. This is the collision problem of RFID systems. As mentioned above, the collision problem can be divided into tag collision and reader collision. Reader collision refers to the fact that the working frequency of the reader overlaps so that the tag cannot select the appropriate reader. When multi-tag corresponds to multi-reader, the communication link between the tag and the reader cannot be established. Tag collision refers to that when multiple tags correspond to a reader, the tag sends data to the reader at the same time, and the signals collide with each other, making the reader unable to obtain relevant information correctly. As readers are active devices, their performance design is powerful, and readers can communicate with each other. Thus, reader collisions can be resolved by communicating between readers and following uniform read rules. It is tag collision that really troubles the RFID system to identify multiple tags at the same time.

The essence of the tag collision problem is channel contention. Collision prevention is to improve the utilization of channels. In order to explain the problem more clearly in principle, this chapter starts from the basic concept of the channel and introduces the basic thinking of solving the problem.

1.2.1 Information Channel and Channel Capacity

Information channel is one of the indispensable parts of signal in a communication system. A channel is a physical medium for transmitting signals from the sender to the receiver, which can be divided into the wired channel and wireless channel. Channel quality affects signal reception and modulation in two ways. On the one hand, when the signal is transmitted in the actual channel, the channel characteristic is not ideal, which will cause the distortion of the signal waveform; On the other hand, there are various noises in the channel, which will affect the signal transmission. Information channels can be divided into additive white Gaussian noise channels, Rayleigh fading channels, and so on [19].

In a broad sense, a channel can be divided into modulation channels and a coding channel according to its functions. The so-called modulation channel refers to the part of the output of the modulator to the input of the demodulator. From the point of view of modulation and demodulation, the modulation channel is all the transformation devices and transmission media from the output of the modulator to the input of the demodulator. Whatever the process, it is simply a transformation of the modulated signal. We only care about the result of the transformation, not the detailed physical process. Therefore, it is convenient to adopt this definition when studying modulation and demodulation.

Similarly, in digital communication systems, if we focus only on the discussion of the coding and decoding, it is beneficial to adopt the concept of coding channel. Coding channel refers to the part from the output of the encoder to the input of the decoder. This is defined because, from the perspective of coding and decoding, the output of the encoder is a sequence of numbers, while the input of the decoder is also a sequence of numbers, which may be different sequences of numbers. Therefore, from the output of the encoder to the input of the decoder, it can be summarized by a box that transforms the sequence of numbers.

In order to analyze the general characteristics of a channel and its influence on signal transmission, the mathematical models of the modulated channel and coded channel are introduced based on the definition of a channel.

First, the modulation channel model is discussed. In any mode of communication with a modulation and demodulation process, the modulated signal output by the modulator is fed into the modulation channel. For studying the performance of modulation and demodulation, we only care about the final result of the modulated signal after it passes through the modulation channel, that is, we only care about the relationship between the output signal and the input signal of the modulation channel, regardless of the transformation of the signal in the modulation channel or the transmission media selected. After a lot of investigation on the modulation channel, it can be found that it has the following commonalities:

(1) There are a pair (or many) of inputs and a pair (or many) of outputs.
(2) Most channels are linear, which satisfies the superposition principle.
(3) The signal has a delay time through the channel, and it is subject to (fixed or time-varying) losses.
(4) Even if there is no signal input, there is still some power output (noise) at the output end of the channel.

According to the above commonalities, we can use a two-pair (or multi-pair) time-varying linear network to represent the modulation channel, which is called the modulation channel model.

For the two-pair channel model, the relation between its output and input is

$$e_0(t) = f[e_i(t)] + n(t) \tag{1.1}$$

where $e_i(t)$ is the input modulated signal; $e_o(t)$ is the total channel output waveform; $n(t)$ is the additive noise (or additive interference). Here $n(t)$ is independent of $e_i(t)$. $f[e_i(t)]$ represents the linear transformation of the modulated signal through the network.

Now, let's say we can write $f[e_i(t)]$ as $k(t)e_i(t)$, where $k(t)$ depends on the properties of the network, and $k(t)$ times $e_i(t)$ reflects what the properties of the network do to $e_i(t)$. The existence of $k(t)$ is a distraction for $e_i(t)$, usually referred to as multiplicative noise. Thus, Eq. (1.1) can be expressed as

$$e_0(t) = k(t)e_i(t) + n(t) \tag{1.2}$$

Equation (1.2) is the mathematical model of a two-pair channel.

From the above analysis, it can be seen that the influence of the channel on the signal can be summarized into two points: one is multiplicative interference $k(t)$, and the other is additive interference $n(t)$. If we understand the characteristics of $k(t)$ and $n(t)$, we can figure out the specific effect of the channel on the signal. The different characteristics of the channel are reflected in the channel model. Only $k(t)$ and $n(t)$ are different.

In general, $k(t)$ is a complex function that may include a variety of linear and nonlinear distortions. At the same time, because the delay and loss characteristics of the channel change randomly at any time, $k(t)$ can only be expressed by a random process. However, a large number of observations indicate $k(t)$ that in some channels remains largely unchanged over time. In other words, the channel's effect on the signal is fixed or changes very slowly. While some channels, otherwise, their $k(t)$ is a random change quickly. Therefore, when analyzing multiplicative interference $k(t)$, channels can be roughly divided into two categories. A class of channels are called constant parameter channels, that is, their $k(t)$ can be regarded as not changing with time or not changing at all. The other is called a random parameter channel, which is a generic term for a non-constant parameter channel, or $k(t)$, which varies rapidly and randomly.

Now, let's talk about the coded channel model. It is obviously different from the modulation channel model. The effect of the modulated channel on the signal is to make the modulated signal undergo analog changes through $k(t)$ and $n(t)$. The effect of the coded channel on the signal is a transformation of the digital sequence, which changes from one digital sequence into another. Therefore, the modulation channel is sometimes regarded as an analog channel and the coded channel as a digital channel.

Since the coded channel contains the modulation channel, it is affected by the modulation channel. However, from an encoding and decoding point of view, this effect is already reflected in the demodulator's sequence of output digits, that is, the output digits will slip with some probability. Obviously, the worse the modulated channel, that is, the less ideal the characteristic and the more serious the additive noise, the greater the probability of error will be. Therefore, the coded channel model can be described by the transfer probability of numbers.

The mathematical model of the two-pair channel is described in detail above, and the following concepts are very important in discussing the effect of the channel on the tag work.

Channel bandwidth is defined as the bandwidth of the channel allowed to pass, referred to as bandwidth [20]. The bandwidth is calculated as follows:

$$BW = f_2 - f_1 \tag{1.3}$$

where f_2 is the highest frequency that the signal can pass in the channel; f_1 is the lowest frequency. Both are determined by the physical properties of the channel. When the composition of the channel is determined, the bandwidth is determined.

According to Shannon calculation, under typical circumstances, that is, under the interference of Gaussian white noise, channel capacity is written as

$$C = BW \log_2\left(1 + \frac{S}{N}\right) \tag{1.4}$$

where the unit of BW is Hertz, S is signal power, and N is noise power, both in watts. Channel capacity is proportional to channel bandwidth, which determines the maximum information transfer rate in the current channel.

It can be seen from the channel definition and channel capacity calculation formula that, in a certain RFID tag reading and writing scene, the background noise is relatively stable, the transmitting power is certain, and the medium condition is certain, so the channel capacity has a fixed upper limit. Therefore, the anti-collision algorithm is to improve the utilization of the channel under the condition of fixed channel capacity.

According to Eq. (1.4), channel capacity is known. The transmitter of the single-input single-output (SISO) system is a single reader and the receiver is a single tag. Thus, the channel matrix H is the identity matrix, and the SNR is ξ. According to Eq. (1.4), normalized channel capacity is

$$C = \log_2(1 + \xi) \tag{1.5}$$

The receiver of the single-input multi-output (SIMO) system is equipped with M tags, and the transmitter has only $N = 1$ reader antenna. Channel matrix $H = [h_1 h_2 \ldots h_M]$, where h_i represents the channel coefficient of the ith root reader antenna from the transmitter to the receiver, then the channel capacity is

$$C = \log_2\left(1 + HH^{\mathrm{T}}\xi\right) = \log_2\left(1 + \sum_{i=1}^{M} |h_i|^2 \xi\right) = \log_2(1 + M\xi) \tag{1.6}$$

The transmitter of the multiple-input single-output (MISO) system is equipped with N reader antennas, while the receiver has only $M = 1$ tag. Channel matrix $H = [h_1 h_2 \ldots h_N]$, where h_j represents the channel coefficient from the jth reader antenna at the transmitting end to the receiving end, then the channel capacity

$$C = \log_2\left(1 + HH^{\mathrm{T}}\xi\right) = \log_2\left(1 + \sum_{j=1}^{N} \left|h_j\right|^2 \xi\right) = \log_2(1 + N\xi) \tag{1.7}$$

The transmitting end of the MIMO system is equipped with multiple reader antennas, and the receiving end is equipped with multiple tags, that is, M and N are larger than 1, and the channel capacity is as Eq. (1.8).

$$C = \log_2(1 + MN\xi) \tag{1.8}$$

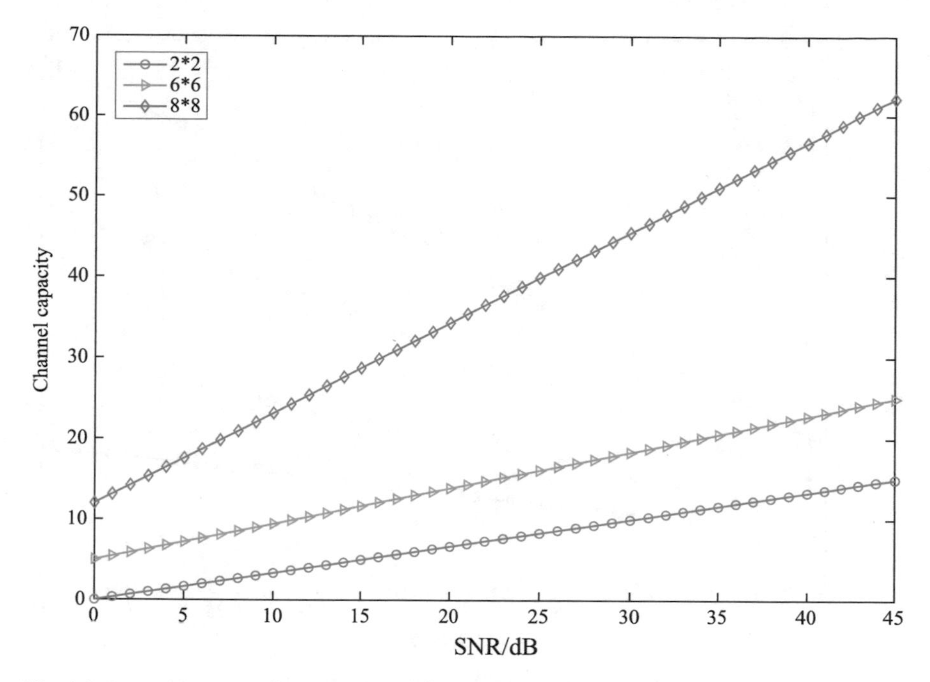

Fig. 1.1 The comparison of the four channel capacities

Compared with SISO, SIMO, and MISO systems, the channel capacity of the MIMO system increases linearly, and the channel capacity of the four systems and SNR ξ is logarithmic. Therefore, it is possible and effective to increase the data transmission rate and channel capacity by increasing the number of transmitting antennas and receiving tags.

MIMO (4 $\times$ 4) and SISO (1 $\times$ 1), SIMO (1X4), and MISO (4 $\times$ 1) channels were selected for simulation, with the minimum SNR_{min} = 0dB and maximum SNR_{max} = 45dB. Figure 1.1 shows the relationship between channel capacity and SNR of SISO, SIMO, MISO, and MIMO channels.

As can be seen from Fig. 1.1, the channel capacity of SIMO (1 $\times$ 4) and MISO (4 $\times$ 1) systems is basically the same; Under the same signal-to-noise ratio, the channel capacity of MIMO (4 $\times$ 4) system is significantly higher than that of SIMO (1 $\times$ 4) and MISO (4 $\times$ 1) system. The channel capacity of SIMO (1 $\times$ 4) and MSO(4 $\times$ systems is increased compared to that of SISO systems.

The MIMO channel capacity of different tag combinations was analyzed when $N = M$, and the 2 $\times$ 2, 4 $\times$ 4, 6 $\times$ 6, and 8 $\times$ 8 multi-tag and multi-antenna systems were selected to establish the simulation model. Figure 1.2 shows the relationship curve between the channel capacity of different tag combinations and SNR. At this point, the channel capacity increases logarithmically with the number of readers and tags. In other words, the channel capacity can be increased exponentially by increasing the number of antennas at the sending end and tags at the receiving end.

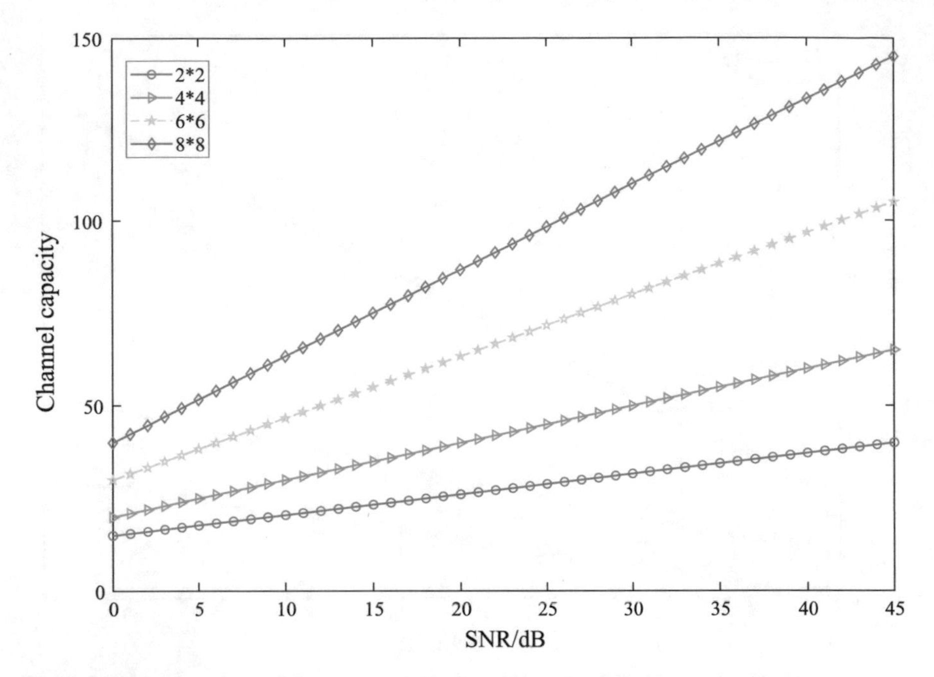

Fig. 1.2 The comparison of the MIMO channel capacities for different tag combinations

It can also be seen from the channel definition and channel capacity calculation formula that, in a certain RFID tag reading and writing scene, the background noise is relatively stable, the transmitting power is certain, and the medium condition is certain, so the channel capacity has a fixed upper limit. Therefore, the anti-collision algorithm is to study how to improve the utilization of channels under the condition of fixed channel capacity.

As a matter of fact, scholars from all over the world have already begun to explore the methods to improve channel utilization and realize multiple access in a communication system to overcome the problem of signal collision [21]. Therefore, it is necessary to make a simple analysis of these methods.

(1) Space-division multiple access, SDMA

The space-division multiple access method is used to identify multiple objects by dividing the space scope. One application for RFID systems is to divide tags and readers by distance, at the expense of having more readers in the space; Another application is to add a phased array antenna to the reader space. Readers use different angles and positions to distinguish different spatial tags. Phased array antennas used for air separation multiple access are complex and costly, and are used only for special occasions such as mass marathons.

(2) Frequency-division multiple access, FDMA

Frequency-division multiple access means that the transmission channel is divided into several different carrier frequency channels, each channel between isolated areas in order to prevent interference between the channel. In RFID systems, reader-to-tag communication, known as downlink, is fixed in frequency. If frequency-division multiple access is applied, the signal input from the tag to the reader, i. e., the uplink, must be transmitted at a variable frequency so that the reader can pass through different tags in the frequency zone. In the RFID system using frequency-division multiple access technology has obvious drawbacks, each tag must have a receiving module, working frequency of the read and write also need to widen, so it will improve the cost.

(3) Time-division multiple access, TDMA

Time-division multiple access (TDMA) is the division of the whole channel transmission information into thousands of time slots, allocated to multiple users. It is characterized by fixed time slot distribution and suitable for transmitting digital information. The application of time-division multiple access is very extensive, the promotion and application of telephone is a classic application example. For RFID systems, time division multiple access is the most widely used basic principle in collision avoidance algorithms, such as the ALOHA method, binary tree method, etc.

(4) Code-division multiple access, CDMA

Code-division multiple access (CDMA) technology has been applied to civil communication as the quality of wireless communication becomes more and more demanding. Code-division multiple access technology can be traced back to spread spectrum communication technology for military use during World War II. When code division multiple access communication is applied, the sender first modulates the information to be transmitted with pseudo-random code, and then sends it with the carrier. The receiver needs the same pseudo-random code to unexpand the received signal. Code-division multiple access (CDMA) is not separated from frequency band transmission, so it can distinguish multi-channel information transmitted at the same time by coding, and discard unwanted signals as noise. Code-division multiple access technology channel capacity is not large, frequency band utilization is not high, address code selection also needs strong computing power, the receiver needs a long time to capture address code. This makes code-division multiple access difficult to apply to RFID systems.

Traditional multiplexing methods in the field of communication are difficult to be directly applied to the problem of RFID multi-tag reading, but with the development of RFID, these methods contain the possibility of solving the problem. In terms of tag collision prevention, considering the complexity and cost of tags, most RFID tags used in practice are passive tags. Therefore, this chapter focuses on the problem of passive tag collision prevention. Collision prevention methods are divided into two ways. One is to realize tag collision prevention by software algorithm according to existing resources. The other is to optimize the hardware environment, including the tag according to a specific geometric distribution, reduce space electromagnetic interference, design of a novel antenna, build a multi-in-multi-out system, and other physical methods to achieve tag collision prevention.

1.2.2 Software Anti-Collision

At present, the most commonly used software anti-collision algorithms can be divided into two categories. One is a probabilistic algorithm represented by the ALOHA algorithm(ISO/IEC 18,000–6 Type A Tags as defined by the standard). The other is a deterministic algorithm represented by a binary tree algorithm(ISO/IEC 18,000–6 Type B Tags as defined by the standard) [22].

The basic idea of the ALOHA algorithm is time-division multiple access. ALOHA algorithm is simple, easy to implement, and suitable for low-cost RFID systems. However, due to the randomness of this kind of algorithm, that is, the possibility that a tag cannot be recognized in a considerable period of time, this kind of method is called the probabilistic method.

(1) Pure ALOHA algorithm

Pure ALOHA algorithm is a simple time-division multiple access method. ALOHA was originally a greeting used by Hawaiians. In the 1970s, the ALOHA system at the University of Hawaii was the first computer system to use radio broadcasting as a communication facility. The system is applied to the ground network using ALOHA protocol, or pure ALOHA. Pure ALOHA is a simple idea, with tags sending data to readers in real time, which creates collisions. However, based on the feedback of the broadcast channel, the sender can carry out collision detection. Once the sender detects a collision, it will wait for a random time to resend the ton until it is successfully sent. The basic process is shown in Fig. 1.3.

This algorithm is suitable for occasions where the real-time requirement is not high. As long as the communication identification process is long enough, the reader can identify all the tags within its reading and writing range. However, this algorithm is only used when a small amount of information is transmitted to the read-only tag of the reader, and there will be a collision of data parts.

(2) Slotted ALOHA algorithm

The time slot ALOHA algorithm was proposed in 1972. Its basic idea is to divide the channel into discrete time slots, and the size of the time slots is determined by the system, but the time slots are at least greater than the time required to send data. It is stipulated that only at the critical point of each time slot, the tag will actively send data to the reader, and the time of sending data is fixed within each time slot,

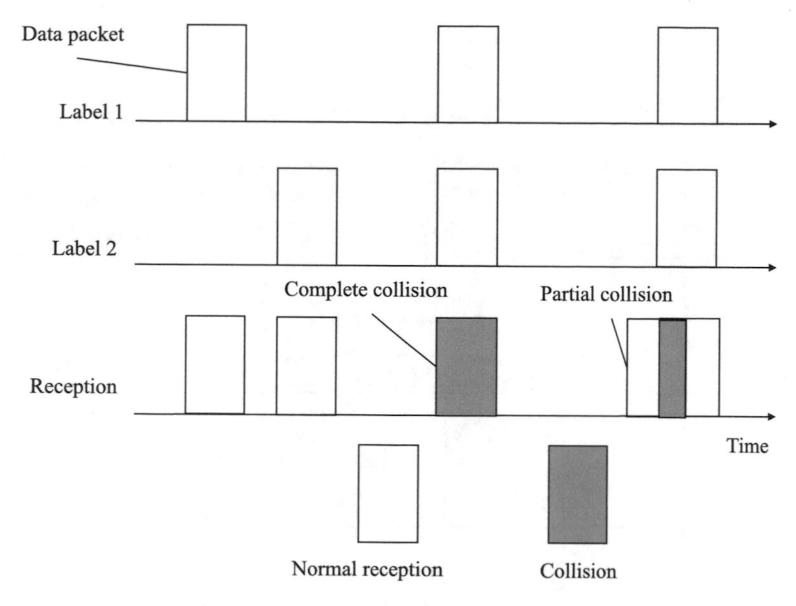

Fig. 1.3 Schematic diagram of pure ALOHA algorithm

so that the data can be sent successfully or completely conflict, waiting for the next time slot to send again. This effectively avoids the problem of partial conflict of data.

The principle of time slot ALOHA is shown in Fig. 1.4, assuming that there are four tags in the scope of the reader. In the first time slot, tag 1 and tag 2 send data to the reader at the same time, the data of the two tags collide, the transmission fails, then tag 1 and tag 2 will be sent again after a certain delay. In the second time slot, tag 3 sends the data. During the process, there is no collision with the data of other tags, so the data is sent successfully. In the third time slot, the data sent again by tag 2 collided with the data sent by tag 4, and failed to be sent. After a certain delay, tag 2 and tag 4 would be sent again. In the fourth slot, the data sent again by tag 1 does not collide with the data of other tags, the data is sent successfully, and so on.

(3) Fixed Framed Slotted ALOHA (FFSA) algorithm

FFSA algorithm is an improvement of the ALOHA algorithm. Based on the time slot ALOHA, the system composed N time slots into a frame. The technical process is shown in Fig. 1.5. At the beginning of the recognition process, the reader sends a command containing the number of slots N to all tags in the recognition field. After receiving the command, reset the time slot counter to 1 and start recording the number of time slots. At the same time, select a number from 1 to N as its sending time barrier value. When the gap counter value reaches the selected value, the tag begins to send a reply message to the reader. If the tag is successfully recognized by the reader, it will exit the system. If two tags within a time slot respond, a collision occurs, and the system waits for the next frame to be read. When one frame is finished, the reader starts another frame with a time slot of N.

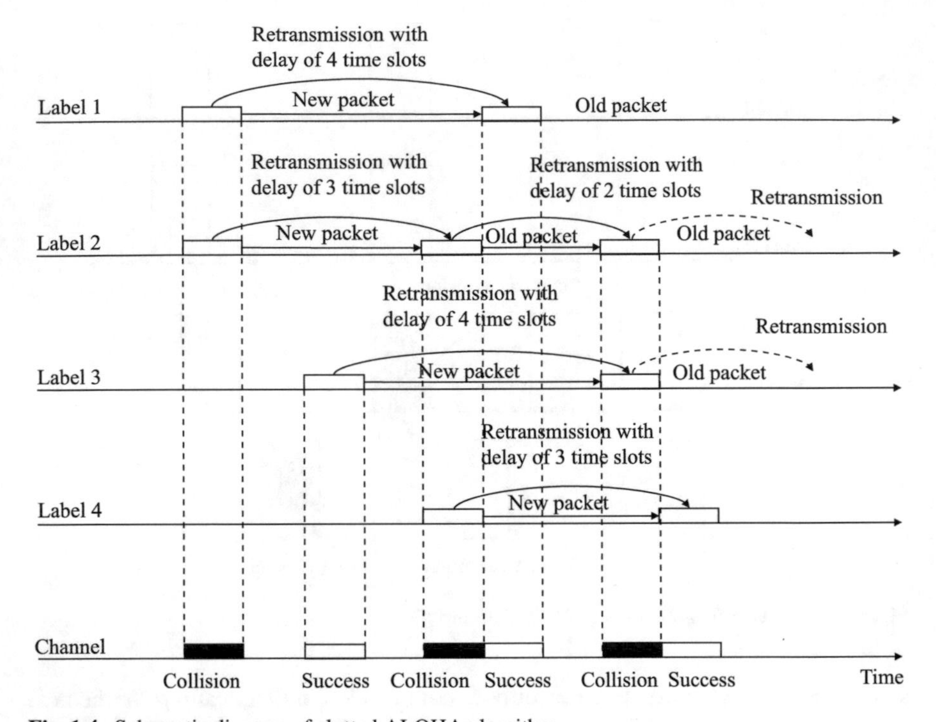

Fig. 1.4 Schematic diagram of slotted ALOHA algorithm

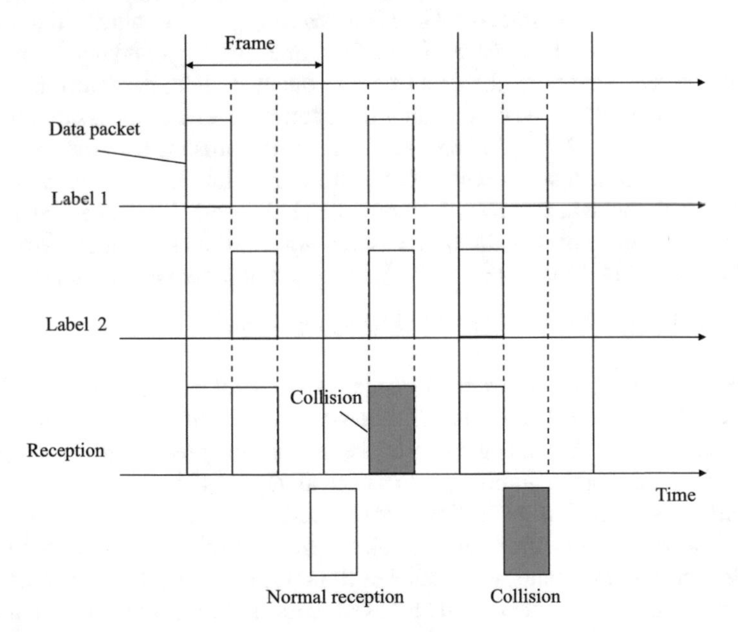

Fig. 1.5 Schematic diagram of fixed frame slotted ALOHA

The disadvantage of the fixed slot ALOHA method is that the system efficiency is very low when the number of slots in the frame is different from the number of tags.

(4) Dynamic framed slotted ALOHA(DFSA) algorithm

The DFSA algorithm is an improvement and supplement to the FFSA method. The basic idea is to dynamically increase or decrease the frame size so that the time gap size matches the number of unrecognized tags within the scope of the reader, thus making the RFID system the most efficient. The process is shown in Fig. 1.6. The key to the DFSA algorithm is to correctly obtain information such as the number of unrecognized tags and the number of slots when collisions occur.

In deterministic algorithms, the most common one is the binary search tree algorithm. The basic idea of the binary search tree algorithm is that the reader sends the request command containing the whole serial number. After the tag group receives the request command, it compares its serial number with it, and returns the data if it meets the requirements. In the event of a collision, the reader separates the tag based on the collision location of the sequence number. A binary search tree algorithm is complicated and takes a long time to identify, but there is no problem that a tag

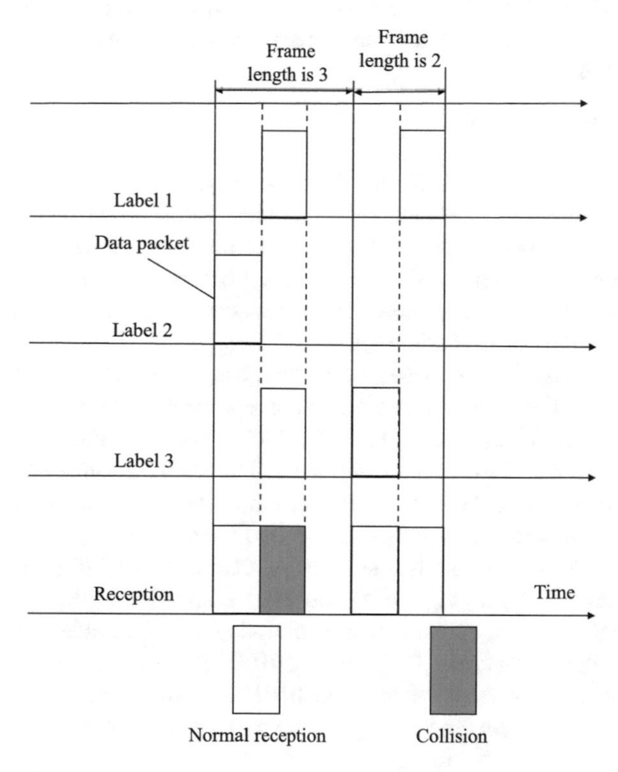

Fig. 1.6 Schematic diagram of dynamic frame slotted ALOHA

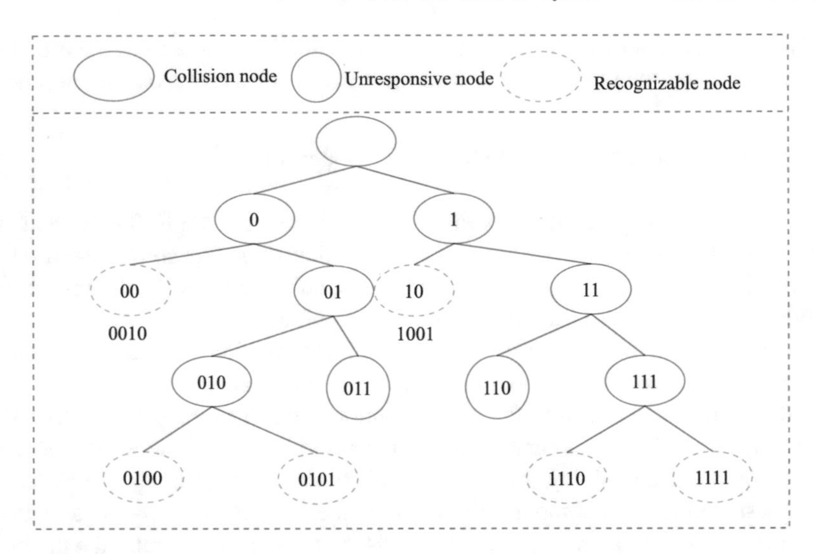

Fig. 1.7 Tree algorithm schematic diagram

cannot be identified for a long time, so it is called the deterministic method. The basic binary search tree, the backward binary search tree, and the dynamic binary table tree are introduced here.

(1) Binary search tree, BST

Binary search tree algorithm is similar to dichotomy, searching by a tree branch. All tag sequence numbers uniquely identified in binary form can form a complete binary tree. The serial number of the tag that synchronously sends signals to the reader within the scope of the reader also constitutes a binary tree. The reader repeatedly screens the branches of the complete binary tree according to the collision of signals, and finally finds the corresponding tags.

As shown in Fig. 1.7, suppose you have six tags with ids of 0010,0100,0101, 1001,1110, 1111. The query starts from the parent node, and the reader sends information 0 to the tag. All tags with ID first 0 respond, and the response signal is sent to the reader, that is, a collision occurs at node 0; The reader again sends a message 00 to the response tag, and only the tag 0010 responds, that is, the tag is recognized; The reader sends the message 01 again, the tags 0100 and 0101 respond, and the reader sends the response signal, that is, the collision occurs at node 01; The reader sends the information 010 again, tags 0100 and 0101 respond, and the reader sends the response signal, that is, a collision occurs at node 010; The reader sends a message 0100 to the response tag again. Only the tag 0100 responds, that is, the tag is recognized; The reader again sends the message 0101 to the response tag, and only the tag 0101 responds, that is, the tag is recognized. At this point, the tag query of node 0 is finished, and the tag query of node 1 is the same.

(2) Regressive-style binary search tree, RBST

After the basic binary search tree algorithm completes a recognition, it loops from the root node to the next reading process. In order to reduce the search times of the algorithm, the regressive-style binary search tree will not go back to the root node for re-identification after the completion of one time of recognition, but will go back to the node of the last collision to continue the query.

The steps of the regressive-style binary tree algorithm are as follows:

(1) The tag enters the working range of the reader, and the reader sends a maximum serial number signal. The serial number of all tags is less than or equal to the maximum serial number, so the tag transponder sends its serial number back to the reader at the same time.
(2) Because of the uniqueness of the tag sequence number, if the number of tags is not less than two, the system will collide. At this point, the starting position of the corresponding collision in the maximum sequence number will be 0. If it is lower than that, then it stays the same. If it is higher than that, then it is 1.
(3) The reader sends the processed maximum sequence number to the tag, and the tag transponder compares its own sequence number to that value and sends back its own sequence number if it is less than or equal to that value.
(4) Loop through the above steps until a transponder with the minimum serial number is selected, and after communicating with it normally, the tag transponder is ordered to sleep, that is, it will not respond to the reader request command unless it is reenergized.
(5) Return to the last node where the collision occurred, obtain the maximum sequence number corresponding to the node, and repeat the above process to identify each tag from small to large.
Following the example above, assuming the tag 0100 is identified in a search, instead of the root node 0 returned by the base binary search tree, the reader returns the collision node 010 and searches for the tag 0101 (the maximum value of the collision node). In this way, the next tag can be immediately identified, improving the identification efficiency.
(6) Dynamic Binary Search Tree, DBST.
The improvement of the dynamic binary search algorithm to basic binary search algorithm focuses on reducing the amount of information transferred to the system. In practical application, the identification serial number of RFID tags may be more than 4 bits, as many as a dozen bytes. For such a tag with a long serial number, if the value of the serial number is transmitted completely every time, the amount of data to be transmitted will be relatively larger and the time will be longer. The dynamic binary search was developed to solve this problem.

The steps of the dynamic binary tree algorithm are as follows:

(1) The tag enters the working range of the reader, and the reader sends a maximum serial number signal. The serial number of all tags is less than or equal to the

maximum serial number, so the tag transponder sends its serial number back to the reader at the same time.

(2) Due to the uniqueness of the tag sequence number, when the number of tags is not less than two, the system will inevitably collide. In case of collision, the reader will set the corresponding starting position of collision in the maximum sequence number to 0, and the position lower than that will remain unchanged.
(3) The reader sends the processed start and low bits of the collision to the tag transponder, which compares its sequence number with the value, equals the value, and sends back the remaining bits of its sequence number.
(4) Loop through the above steps until a tag with the minimum sequence number is selected. After the reader communicates normally with the tag, a command is issued to put the transponder into hibernation, that is, it will not respond to the reader request command unless it is reenergized.
(5) Repeat until each tag is identified in sequence from small to large.

1.2.3 Physical Anti-Collision

In addition to using algorithms to avoid tag collisions, the book also proposes new ideas for reducing tag collisions through physical methods. Here are four ways to do this.

(1) Optimal geometric distribution of RFID multi-tag based on Fisher information matrix

Using software algorithm can only prevent tag collision, but can not improve the system's reading performance. The Fisher information matrix contains information such as the position and detection value of each tag. By analyzing and calculating the determinant of the Fisher information matrix, the relationship between the tag geometric distribution and the reader can be obtained to obtain the optimal geometric distribution of the multi-tag system. Fisher information matrix is introduced to study the location distribution of RFID tags, and then the optimal geometric arrangement of tags can effectively improve the dynamic performance of the multi-tag system and reduce the reading error.

(2) A method to enhance the system's ability to read multiple tags by adjusting the antenna design

All antennas (except the ideal point source antenna) radiate directionally in space. By changing the antenna manufacturing technology, changing the antenna aspect ratio, increasing the antenna downdip Angle, and so on, adjusting the antenna's direction and resonance frequency can effectively improve the system's performance in multi-tag reading.

(3) The method of constructing RFID multi-input multi-output (RFID-MIMO) system

With the in-depth research on RFID and the rapid development of multi-input multi-output (MIMO) communication, the RFID-MIMO system established by the integration of RFID and MIMO communication presents a broad space for development. MIMO technology improves the reliability of the RFID system by eliminating interference through near-field spatial multiplexing and far-field spatial diversity. MIMO channel has multiple links working at the same frequency, which can lengthen the read–write distance of RFID, reduce the system error rate of RFID and improve the read–write rate of tags without increasing the signal bandwidth.

(4) Methods to reduce the electromagnetic interference in ambient space

It can be seen from Eq. (1.4) that noise in space will have an impact on the communication channel of the RFID system. Therefore, when designing and applying RFID systems, if engineers can accurately grasp the impact of the environment on the electromagnetic field, and correctly use these parameters to optimize the system, then it will be very helpful to improve the whole RFID system's reading performance and reduce the error rate of reading.

For the parameters of the environmental electromagnetic field, the following three points are mainly considered:

(1) The working wavelength of an electromagnetic wave in a medium

$$\lambda = \frac{\sqrt{2}}{f\sqrt{\mu\varepsilon}}\left(\sqrt{1+\left(\frac{\sigma}{\omega c}\right)^2}-1\right)^{-\frac{1}{2}} \tag{1.9}$$

where λ represents the wavelength of the electromagnetic wave in the medium; f is the electromagnetic frequency; ω is the angular frequency, c is the speed of light in a vacuum, they represent the parameters of the electromagnetic wave. μ is the permeability of the medium; ε is the dielectric constant; σ is the conductivity, and these three terms are the parameters that characterize the material.

In this case, the wavelength has an effect on the size of the antenna, so in practical applications, either the antenna design is changed to offset the environmental impact, or the environmental medium is changed to accommodate the antenna.

(2) Attenuation coefficient of dielectric material

$$\alpha = \frac{\omega\sqrt{\mu\varepsilon}}{\sqrt{2}}\left(\sqrt{1+\left(\frac{\sigma}{\omega c}\right)^2}-1\right)^{\frac{1}{2}} \tag{1.10}$$

where α represents the attenuation coefficient of the amplitude of electromagnetic wave in the medium, and the tangent value of α is often used as the dissipation factor in engineering. It can be seen that only when the conductivity is 0, the electromagnetic wave will be lossless propagation in the medium, which does not exist in the practical application. Similarly, the attenuation factor can significantly affect the RFID system's reading distance, which must be considered when deploying the system.

In addition, when the antenna captures space electromagnetic waves, the attenuation coefficient of the antenna base material will also affect the decoding of the system, thus causing obstacles to the reading. Therefore, when designing the antenna, the lower the conductivity of the selected material should be, the better.

(3) Metamaterials and their electromagnetic properties

Metamaterial is a composite material that has an artificially designed structure and exhibits supernormal physical properties not found in natural materials. In this medium, the strength of the electric field, the magnetic field and the electromagnetic wave vector comply with the left-hand rule, hence the term "left-handed material".

When electromagnetic waves travel through metamaterials, they exhibit some bizarre properties:

(1) The group velocity direction of the electromagnetic wave is parallel to the phase velocity direction, that is, the direction of the wave vector is opposite to the propagation direction of energy, and the left-hand law is satisfied between electric field, magnetic field, and wave vector.
(2) Reversed Doppler effect. The observed change in frequency in the left-handed material is the opposite of the effect in the right-handed material. In the right-handed material, when the observer moves toward the source, the observer measures a higher frequency than the source vibrates, which is called the Doppler effect. In the left-handed material, similarly, when the observer is moving toward the source of the wave, the frequency measured by the observer is lower than that of the source of the wave, which is the inverse Doppler effect.
(3) Reversed Snell refraction. The refractive index is negative. At the interface between left-handed material and right-handed material, the refractive rays and incoming rays are on the same side of the normal line. So you have what's called a perfect lens.
(4) Reversed Cerenkov radiation. When a charged particle moves in a medium, an induced current is generated in the medium. By these induced currents, secondary waves are excited. When the velocity of a charged particle exceeds the speed of light in the medium, these secondary waves interfere with the original electromagnetic field and can form a radiation electromagnetic wave. This radiation is called Cerenkov radiation. In the right-handed material, the electromagnetic-excited radiation scatters forward at an acute angle. In the left-handed material, the radiation direction of the electromagnetic wave changes, scattering backward at an obtuse ngle.

1.3 Physical Anti-Collision

1.3.1 Definition of Physical Collision Avoidance

The "physical anti-collision" proposed in this book is relative to the software anti-collision. Research has found that multi-tag collisions that occur in the frequency bands above UHF have both inherent algorithm design flaws and some tags that cannot be identified due to various external physical interferences. Physical anti-collision mainly solves the problem of low batch recognition success rate caused by the latter. Therefore, physical collision avoidance is defined as the use of physical means to solve the problem of multi-tag collision caused by non-software factors.

1.3.2 The Main Features of Physical Collision Avoidance

(1) Reflecting physicality in principle: The main technical methods of physical collision prevention are based on physical principles such as radiophysics, thermodynamic analysis, and electromagnetic analysis.
(2) The verification method reflects the physicality: the verification of physical anti-collision mainly adopts the semi-physical verification method, that is, the physical quantity sensor is used to simulate the actual signal to verify the anti-collision performance of multiple tags in actual application scenarios.
(3) The realization of anti-collision reflects the physicality: the specific realization of physical anti-collision mainly adopts the method of physical space optimization, predicts the optimal physical space distribution of multiple tags through artificial intelligence, and finally optimizes the physical space distribution to maximize the resistance to the outside Physical interference.

1.3.3 The Structure of the Physical Anti-Collision System

The front end of the physical anti-collision system uses physical means to collect data, for example, image sensors are used to collect the geometric characteristics of multi-tag distribution, and the back-end uses neural network algorithms to learn, train, and predict the physical optimal distribution structure, and then adjust the tag arrangement through physical means, angle, to achieve the overall optimal recognition performance of the tag group.

References

1. Xu DL, He W, Li S (2014) Internet of Things in Industries: A Survey. IEEE Trans Ind Inf 10(4):2233–2243
2. Biji KB, Ravishankar CN, Mohan CO et al (2015) Smart packaging systems for food applications: a review. J Food Sci Technol 52(10):6125–6135
3. Valero E, Adán A, Cerrada C (2015) Evolution of RFID applications in construction: a literature review. Sensors 15(7):15988–16008
4. Zayou R, Besbe MA, Hamam H (2014) Agricultural and environmental applications of RFID technology. Int J Agricult Environ Infor Syst 5(2):50–65
5. Liukkonen M (2015) RFID technology in manufacturing and supply chain. Int J Comput Integr Manuf 28(8):861–880
6. Griffin JD, Durgin GD, Haldi A et al (2006) RF tag antenna performance on various materials using radio link budgets. IEEE Antennas Wirel Propag Lett 5:247–250
7. Bekkali A, Zou S, Kadri A et al (2015) Performance analysis of passive UHF RFID systems under cascaded fading channels and interference effects. IEEE Trans Wireless Commun 14(3):1421–1433
8. Feng KT, Chen CL, Chen CH (2008) GALE: An enhanced geometry-assisted location estimation algorithm for NLOS environments. IEEE Trans Mob Comput 7(2):199–213
9. Liu R, Huskic G, Zell A (2015) On tracking dynamic objects with long range passive UHF RFID using a mobile robot. Int J Distrib Sens Netw 2015:1–12
10. Feng C, Zhang W, Li L et al (2015) Angle-based chipless RFID tag with high capacity and insensitivity to polarization. IEEE Trans Antennas Propag 63(4):1789–1797
11. Luh YP, Liu YC (2013) Measurement of effective reading distance of UHF RFID passive tags. Modern Mech Eng 3:115–120
12. Van Der Togt R, Van Lieshout EJ, Hensbroek R et al (2008) Electromagnetic interference from radio frequency identification inducing potentially hazardous incidents in critical care medical equipment. J Am Med Assoc 299(24):2884–2890
13. Coustasse A, Meadows P, Hibner T (2015) Utilizing radiofrequency identification technology to improve safety and management of blood bank supply chains. Telemedicine and e-Health 21(11):938–945
14. Oliveira RR, Cardoso IMG, Barbosa JLV et al (2015) An intelligent model for logistics management based on geofencing algorithms and RFID technology. Expert Syst Appl 42:6082 6097
15. Zhu L, Yum TS (2011) A critical survey and analysis of RFID anti-collision mechanisms. IEEE Commun Mag 5:214–221
16. Zhu L, Yum TS (2010) The optimal reading strategy for EPC Gen-2 RFID anti-collision systems. IEEE Trans Commun 58(9):2725–2733
17. von Pidoll U (2009) An overview of standards concerning unwanted electrostatic discharges. J Electrostat 67:445–452
18. Fescioglu-Unver N, Choi S H, Sheen D et al (2015) RFID in production and service systems: technology. applications and issues. Infor Syst Frontiers 17(6):1369–1380
19. Castro L, Lefebvre E, Lefebvre LA (2013) Adding intelligence to mobile asset management in hospitals: the true value of RFID. J Med Syst 37(5):1–17
20. Yusoff AN, Abdullah MH, Ahmad SH et al (2002) Electromagnetic and absorption properties of some microwave absorbers. J Appl Phys 92(2):876–882
21. Gibaldi A, Canessa A, Solari F et al (2015) Autonomous learning of disparity-vergence behavior through distributed coding and population reward: basic mechanisms and real-world conditioning on a robot stereo head. Robot Autonomous Syst 71:23–34
22. Righetti L, Kalakrishnan M, Pastor P et al (2014) An autonomous manipulation system based on force control and optimization. Auton Robot 36(1–2):11–30
23. Pinto AM, Moreira AP, Correia MV et al (2014) A flow-based motion perception technique for an autonomous robot system. J Intell Rob Syst 75(3–4):475–492

24. Kruse T, Pandey AK, Alami R et al (2013) Human-aware robot navigation: a survey. Robot Autonomous Syst 61(12):1726–1743
25. Zhang Z, Yue S, Zhang G (2015) Fly visual system inspired artificial neural network for collision detection. Neurocomputing 153:221–234
26. Song C, Xie S, Zhou Z et al (2015) Modeling of pneumatic artificial muscle using a hybrid artificial neural network approach. Mechatronics 31:124–131
27. Mano M, Capi G, Tanaka N et al (2013) An artificial neural network based robot controller that uses rat's brain signals. Robotics 2(2):54–65
28. Nguyen HN, Zhou J, Kang HJ (2015) A calibration method for enhancing robot accuracy through integration of an extended Kalman filter algorithm and an artificial neural network. Neurocomputing 151:996–1005
29. Soltanpour MR, Khooban MH (2013) A particle swarm optimization approach for fuzzy sliding mode control for tracking the robot manipulator. Nonlinear Dyn 74(1–2):467–478
30. Fateh MM, Azargoshasb S (2014) Discrete adaptive fuzzy control for asymptotic tracking of robotic manipulators. Nonlinear Dyn 78(3):2195–2204
31. Zhou Q, Li H, Shi P (2015) Decentralized adaptive fuzzy tracking control for robot finger dynamics. IEEE Trans Fuzzy Syst 23(3):501–510
32. Ferrauto T, Parisi D, Di Stefano G et al (2013) Different genetic algorithms and the evolution of specialization: a study with groups of simulated neural robots. Artif Life 19(2):221–253
33. Arkin RC (1998) Behaviour Based Robotics. MIT Press, Cambridge, USA
34. Verschure P, Voegtlin T, Douglas RJ (2003) Environmentally mediated synergy between perception and behaviour in mobile robots. Nature 425(6958):620–624
35. Verschure P, Althaus P (2003) A real-world rational agent: Unifying old and new AI. Cogn Sci 27(4):561–590
36. Fuster JM (2003) Cortex and Mind: Unifying Cognition. Oxford University Press, Oxford
37. Freeman WJ (1987) Simulation of chaotic EEG patterns with a dynamic model of the olfactory system. Biol Cybern 56(2–3):139–150
38. Freeman WJ (1991) The physiology of perception. Sci Am 264(2):78–85
39. Freeman WJ (2003) A neurobiological theory of meaning in perception, Part I: Information and meaning in nonconvergent and nonlocal brain dynamics. Int J Bifur Chaos 13(9):2493–2511
40. Freeman WJ (2004) How and why brains create meaning from sensory information. Int J Bifur Chaos 14(2):515–530
41. Harter D, Kozma R (2005) Chaotic neurodynamics for autonomous agents. IEEE Trans Neural Netw 16(3):565–579
42. Islam M, Murase K (2005) Chaotic dynamics of a behavior-based miniature mobile robot: Effects of environment and control structure. Neural Netw 18(2):123–144
43. Arena P, De Fiore S, Fortuna L et al (2008) Perception-action map learning in controlled multiscroll systems applied to robot navigation. Chaos, 18(4), No.043119
44. Kozma R, Freeman WJ (2009) The KIV model of intentional dynamics and decision making. Neural Netw 22(3):277–285
45. Ohl FW, Scheich H, Freeman WJ (2001) Change in pattern of ongoing cortical activity with auditory category learning. Nature 412(6848):733–736
46. Webb B (2002) Robots in invertebrate neuroscience. Nature 417(6886):259–363
47. Pennisi E (2007) Behavior - robot cockroach tests insect decision-making behavior. Science 318(5853):1055
48. Yu X, Sun Y, Liu J et al (2009) Autonomous navigation for unmanned aerial vehicles based on chaotic bionics theory. J Bionic Eng 6(3):270–279
49. Yu X, Sun Y, Liu J et al (2009) Autonomous guidance for intelligent missile based on chaotic perception-action dynamics. J Aerosp Eng 223(G7):853–862
50. Yu X, Sun Y, Liu J et al (2010) Autonomous spatial orientation of robots using chaotic cognition and geometric cues. J Syst Control Eng 224(I2):139–152
51. Yu X, Yu H (2011) A novel low-altitude reconnaissance strategy for smart uavs: Active perception and chaotic navigation. Trans Instit Meas Control 33(5):610–630

52. Yu X, Zhao Z (2011) A Novel Reactive Navigation Strategy for Mobile Robots Based on Chaotic Exploration and TTM Self-construction. COMPEL-The Int J Comput Math Electr Electron Eng 30(2):590–602
53. Li H, Zhao Z, Yu X (2012) Grey theory applied in non-subsampled contourlet transform. IET Image Proc 6(3):264–272
54. Yu X, Wang D, Zhao Z (2019) Optimal distribution and semi-physical verification of RFID multi-tag performance based on image processing. In: Semi-physical verification technology for dynamic performance of internet of things system
55. Liu Q, Zhao Z, Li Y et al (2012) Feature selection based on sensitivity analysis of fuzzy ISODATA. Neurocomputing 85:29–37
56. Liu Q, Zhao Z, Yu X et al (2013) A novel method of feature selection based on SVM. J Comput 8(8):2144–2149
57. Zhao Z, Liu L, Li Y et al (2014) Application of holographic double exposure interferometry in the displacement measurement of a cantilever beam based on digital image processing. Lasers Eng 28(1):81–94
58. Shen L, Zhao Z, Zhu X et al (2015) A design for a remote condition monitoring system for an optical fibre smart structure based on advanced reduced instruction set computing (RISC) machines (ARM) and general packet radio service (GPRS). Lasers Eng 30(1–2):15–29
59. Shen L, Zhao Z, Chen M et al (2015) A novel method of damage model recognition for intelligent composite structures based on double-fiber sensors network. Optik 126(21):3295–3298
60. Yu Y, Yu X, Zhao Z et al (2016) Measurement uncertainty limit analysis of biased estimators in RFID multiple tags system. IET Sci Meas Technol 10(5):449–455
61. Yu Y, Yu X, Zhao Z et al (2016) Online measurement of alcohol concentration based on radio frequency identification. J Test Eval 44(6):2077–2084
62. Yu Y, Yu X, Zhao Z et al (2017) Influence of temperature on the dynamic reading performance of UHF RFID system: theory and experimentation. J Test Eval 45(5):20150466

Chapter 2
RFID System Physical Anti-Collision Experimental Verification

Radio-frequency identification (RFID) systems are widely used because of their accurate identifications, large amount of information storages, and low costs. At the same time, the use of multi-tag is also more and more universal. The function of RFID tag is similar to the barcode, but RFID tag has more advantages in many applications: RFID tags have tiny communication and computation power, among them can tagged objects intelligent to some extent [1]. Some tags because of their small size, ruggedness, and other advantages can be applied to the irregular surface to achieve the purpose of detection [2].

The sensitivity of the tag has a significant impact on the performance of the tag, and it cannot be ignored. The sensitivity of the tag-related research is also carried out in varying degrees. Wang proposed for handheld radio-frequency identification (RFID) reader which can effectively improve the sensitivity of the reader [3]. Petrariu used a longer reading distance to obtain an additional antenna RFID chip, which provides a reading sensitivity of –20.5 dBm [4]. Boaventura proposed the measurement of sensitivity improvement in passive RFID chips when interrogated by a custom-built RFID reader with improved powering waveforms [5]. The research on the sensitivity of single tag has been studied in different degrees and different aspects, but the research on the sensitivity of the tags group has not progressed much.

In this paper, the image matching technique is used to locate the tags and find out the influence of the relative position of the tags on the sensitivity of the tag group. In this paper, SIFT algorithm, SURF algorithm, and ORB algorithm are, respectively, used for image matching test. Three algorithms are compared and analyzed, respectively, from the execution time of the algorithm, the number of feature points and the number of correct matching points, and the SURF algorithm most suitable for image feature matching is selected. Then the algorithm is used to locate the key tags in the tag group.

The performance of RFID tags can be obtained from many aspects, sensitivity is one of its important applications. This paper is mainly divided into two parts. The first part is the experimental study of RFID tag group performance detection, and the second part is the study of tag distribution based on image matching. In

X. Yu et al., *Physical Anti-Collision in RFID Systems*,
https://doi.org/10.1007/978-981-16-0835-3_2

the first part of this paper, the definition of tag group sensitivity is described and a method for measuring tag group sensitivity is proposed. Secondly, a dynamic testing system based on photoelectric sensing is designed based on the practical application scenarios and measurement needs. Finally, the structure and number of tags in tag group were changed, and the influence of key tag positions on the sensitivity of tag group was studied. In the second part, according to the built RFID tag group dynamic test system, aiming at the impact of the tag group sensitivity due to the different arrangement positions of tags in the batch reading of RFID tags, the tag matching and positioning is carried out in combination with the image processing method to find the impact of the relative position of tags on the tag group sensitivity. This study is of great significance to the evaluation and optimization of RFID multi-tag performance and provides an important basis for upgrading the sensitivity measurement system of tag group to the performance optimization analysis system.

2.1 RFID Tag Group Performance Measurement

2.1.1 Tag Group Sensitivity

The test items defined in the RFID tag performance test specification mainly include two categories: static performance test and dynamic performance test. The dynamic performance test is a sports-oriented test item and it can reflect the performance index of the tag in the motion identification application. According to the international standard EPC global, the unipolar radiated power received by the tag is expressed as the tag sensitivity in the dynamic test of the tag.

Friis Transmission formula is one of the most important antenna theories. It relates transmission power, antenna gain, distance, wavelength, and received power. The Friis formula is used to calculate the received power from one antenna to the second, also known as the power transfer equation, as shown in (2.1).

$$r = \frac{\lambda}{4\pi}\sqrt{\frac{P_t G_t G_t \tau}{P_r}} \tag{2.1}$$

where r is the distance of the sensor to the tag, also known as read distance, P_t is the transmitting antenna power, G_t is the transmitting antenna gain, G_r is the receiving antenna gain, P_r is the receiving power, λ is the working wavelength, and τ is the matching coefficient of tag and antenna. According to Friis formula, we can intuitively get the tag power expression, as shown in (2.2).

$$P_r = \frac{\lambda^2 P_t G_t G_r \tau}{(4\pi)^2} \times \frac{1}{r^2} \tag{2.2}$$

However, for tags group with different performance tags, the sensitivity is not just related to the reading distance. Consider the tag is read in turn, so P_r should be greater than or equal to the sensitivity of each tag. If we do not consider the mutual interference between tags, the value of the group tags sensitivity is equal to the sensitivity of the single tag with the worst performance. Otherwise, the antennas provide more energy to activate the tags group. So, the final sensitivity formula should be in the form of an inequality, as shown in (2.3).

$$P_r' \geq \max\{P_1, P_2, P_3, \ldots, P_i, \ldots, P_n\} \tag{2.3}$$

where n is the number of tags,$i = 1, 2, 3, \ldots, n$. In combination with (2.2), the sensitivity formula of group tags can be expressed as (2.4)

$$P_r' \geq \frac{\lambda^2 P_t' G_t' G_r' \tau'}{(4\pi)^2} \times \frac{1}{r^2} \tag{2.4}$$

where r' is the distance from the sensor to the tags' geometric center when all tags in the tags group are been read, λ is the working wavelength of the tag, P_t' is the power of RFID reader, G_t' is the gain of RFID reader antenna, G_r' is the gain of RFID tags group antenna and $G_r' = \max\{G_i\}, i = 1, 2, 3, \ldots, n$, P_r' is the threshold power of RFID tags group, τ' is the matching coefficient of tag and antenna. If the tags are arranged in a reasonable way, the mutual interference is tiny. Hence, P_r' is approximately equal to the maximum sensitivity of all single tags. In practice, the tags are close and the mutual interference certainly exists. Therefore, P_r' is determined inversely according to the actual distance.

2.1.2 Measurement System Designed

(1) System overview

Generally, in order to measure the sensitivity of RFID tags, the experiment is often conducted in a dark room or a field with less interference, which requires a higher experimental environment. In order to better measure the performance of tag groups, Niklas designed a performance test system for measuring the reading distance and environmental impact of RFID tags in the space, but the system can only carry out limited motion measurement and the test distance is relatively short [6]. Literature [7] designed a test system with a working range of 800–1000 mHz frequency. The system has a wide working range and can complete the measurement of tag sensitivity, but the system power is fixed, which is not conducive to the performance test of tag group. In the literature [8], RFID technology is combined with the warehouse automation transmission system to design an RFID device. This system can fully meet the needs of large warehouse logistics enterprises, but the cost is high, which is not suitable for the laboratory research system.

In order to explore the overall performance of the tag group in the process of movement, the test system needs to complete the task of obtaining the recognition distance of the tag group in the process of movement and collecting the distribution image of the tag. With the system of goods entering and leaving the warehouse as the model, RFID tags are attached to the goods. When the goods enter and leave the warehouse, the system collects the goods information in real time. Therefore, the design of the test system needs to simulate the process of goods in and out of the warehouse, data collection and information processing. In the system designed in this paper, the reader can transmit power-adjustable range, can complete long-distance dynamic test, and can collect the tag image, the cost is low and easy to disassemble, convenient for the use of the tag group performance expansion research. This system makes up for the deficiency of the previous test system, the function is more advanced and perfect.

The dynamic test system designed in this study includes reader antenna, laser rangefinder, CCD camera, drive motor, RFID reader, reflector, guide rail, optical lifting platform, data display, and computer. The system structure is shown in Fig. 2.1.

In Fig. 2.1, the guide rail and driving motor are used to simulate the process of entering and leaving the warehouse of goods, and the reflector plate is used to simulate the surface of goods. The reader antenna, RFID reader, laser rangefinder, and CCD camera are used to simulate the data acquisition module. Computer and data display for data processing. The reader antenna, laser rangefinder, and CCD camera are located at one end of the guide rail and can be set horizontally by an optical lifting platform. Among them, the reader antenna is used to transmit and receive microwave

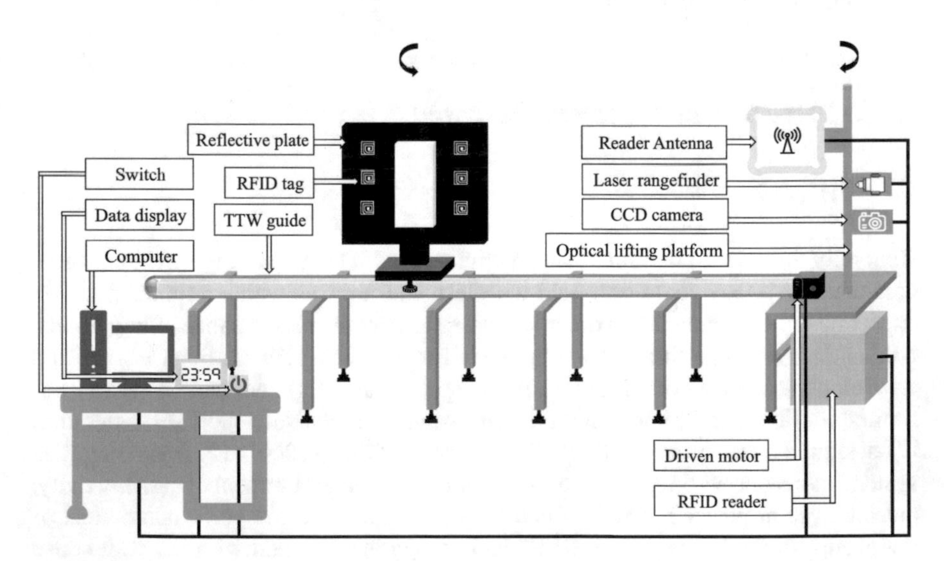

Fig. 2.1 System structure diagram

signals, the laser rangefinder USES the pulse method for ranging, which is used to obtain the maximum distance between the tag and the reader antenna when the tag is read, and the CCD camera is used to collect the tag distribution image of the tag group. The reflector plate simulates the surface of goods, is placed on the guide rail, and is controlled by the driving motor to move back and forth on the guide rail. The RFID tag is attached to the reflector board. When the tag enters the working area of the reader antenna, the reader immediately sends a signal to the laser rangefinder after detecting the signal of the RFID tag. The laser beam emitted by the laser rangefinder falls on the reflector board and displays the measured distance data on the display. Guide rail motion control switch in the display, you can control the beginning and end of the car movement, as well as the car speed and direction of movement. After the measurement is completed, the transmission power and the reading distance of the system are recorded, the trolley is controlled to move to the furthest point of the guide rail, and the measurement is started again.

Because the reading distance of the tag is obtained by the laser rangefinder, and the measurement result of the laser rangefinder is affected by the laser sensor precision in the rangefinder, so the measurement result can be more accurate by improving the range, sensitivity, sampling frequency, and resolution of the sensor. In this experiment, the reading distance between the tag and the tag group was measured for many times, and the average value of the reading distance was substituted into the sensitivity formula of the tag group to reduce the influence of the error of the laser rangefinder on the accuracy of the results.

The experimental test equipments are shown as follows:

(1) Laser ranging sensor: measuring range of 20 m, output frequency of 100 Hz, resolution of 1 cm.
(2) Motion guide: the length of the guide is 15 m, and the motion speed is manually selected from the range of 0.5, 1, and 1.5 m/s.
(3) Reader: the supported frequency range is 840–845 and 920–925 MHz, and the frequency can be fixed. The output power is adjustable from 10 to 30 dBm.
(4) Antenna: the frequency range supported is 840–845 and 920–925 MHz.
(5) Tag: Alien UHF line-polarized tag.
(6) CCD camera: 5 million pixels, the imaging size is 1/3 inch, the exposure time is 0.0063–4615 ms, and the pixel size is 3.75 m $\times$ 3.75 m.

The test environments are shown as follows:

(1) Temperature: the test environment temperature is 20–30 °C.
(2) Relative humidity: the relative humidity of the test environment is 25–75%.
(3) Interference: no obvious interference sources such as electromagnetic, metal, and liquid.

The physical picture of the system is shown in Fig. 2.2.

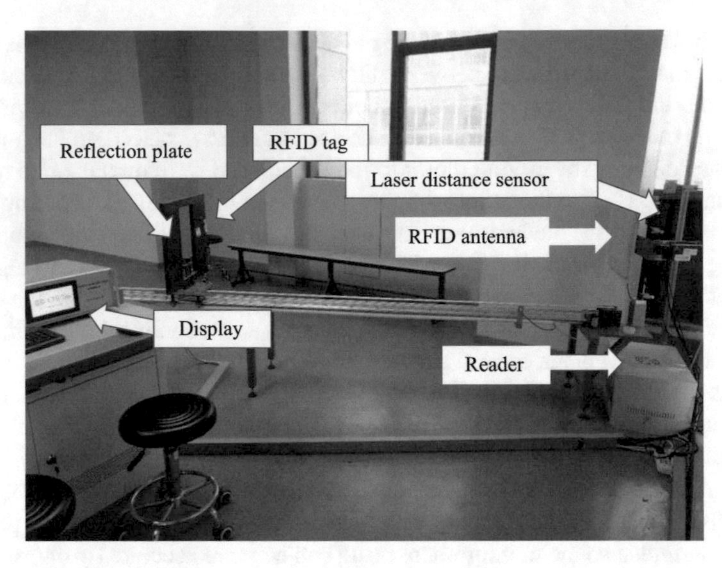

Fig. 2.2 System physical drawing

(2) Test method

When the system is built in the laboratory and the ambient temperature, relative humidity and no obvious interference source meet the requirements, turn on the power and prepare for the test. The steps of specific test method are shown as follows:

(1) Instrument alignment: a reflector plate is placed on the guide rail, and the laser ranging sensor position is adjusted, so that the laser beam emitted by the rangefinder can fall vertically on the reflector plate, adjust the horizontal position of the CCD camera, so that the camera can collect a complete image of tag arrangement.
(2) Tag placement: attach the RFID tag required by the experiment to the reflector board, and ensure that the tag and the reflector board fully fit.
(3) Set the experimental conditions: open the software setting part of the test system, connect the antenna, and set the transmission power of the reader (10−30 dBm), the motion speed of the reflector (0.5, 1, 1.5 m/s), the number of tags (set according to the actual number of tags), the system connection, and confirm the parameters.
(4) Test: click to start measurement, start the drive motor, and the reflector plate starts to move on the guide rail. When the reader detects the tag information, the laser ranging sensor starts to measure, and the result is visible on the display screen. Meanwhile, the CCD completes the acquisition of the tag image.

2.1.3 *Single Tag Sensitivity Measurement*

In order to obtain the performance of the tags in the real environment, this paper proposes a test scheme based on the commonly used UHF RFID tags. This study takes the sensitivity of the tags in the commonly used scenarios as the test target. The test plan is mainly designed as follows:

(1) Modeling: an experimental model is established for the actual application scenario, as shown in Fig. 2.2. A test system is designed according to the application scenario.
(2) Preprocessing of original data: the reading distance of the tag is obtained through the RFID dynamic test platform based on photoelectric sensor, and the data records in the experiment process are sorted out. This step becomes the preprocessing of the original data.
(3) Data analysis: according to the tag group sensitivity defined in Sect. 2.2.1, the sorted original data were substituted into the sensitivity formula for calculation, and the environmental error was calibrated with the sensitivity of a single tag to calculate the value of tag group sensitivity.
(4) Conclusion: the main factors affecting the sensitivity of tag group were analyzed according to the experimental data.

(1) Test method

(1) Instrument alignment: a reflector plate is placed on the guide rail, and the laser ranging sensor position is adjusted, so that the laser beam emitted by the rangefinder can fall vertically on the reflector plate, adjust the horizontal position of the CCD camera, so that the camera can collect a complete image of tag arrangement.
(2) Tag placement: attach the RFID tag used in the experiment to the reflector board, and ensure that the tag and the reflector board fully fit.
(3) Set the experimental conditions: open the software setting part of the test system, connect the antenna, and set the transmitter power of the reader to 24 dBm, the motion speed of the reflector to 1 m/s, and the number of tags to 1. After all options are set, click parameter confirmation.
(4) Test: click to start the measurement, start the drive motor, and the reflector plate starts to move on the guide rail. When the reader detects the tag information, the laser starts to measure, and the ranging result is displayed on the display screen, while the CCD completes the collection of the tag image.

(2) Reading distance test

Five kinds of line-polarized tags with different antenna distribution were selected, and each tag was tested five times in the same test environment, and the average value of the measurement results was used as the reference result of the reading distance. In this measurement, the reader power is set to 24 dBm to ensure that each tag is read in the working range. Because the reading distance of the tag can be used as an index to judge the performance of the tag such as sensitivity, the performance of the five tags is quite different from the measurement results, and the relationship between the sensitivity of the tag group and the performance of the single tag group can be studied as a group of experimental tags. According to the measurement results, the five tags were numbered successively. The tag with the smallest reading distance was No. 1, and the maximum number was No. 5. Among them, the No.1 tag has the worst performance, and it is taken as the key tag in this experiment. The tags selected in the experiment are shown in Fig. 2.3, and the measurement results of single tag are shown in Table 2.1.

(3) Reading distance test

According to Eq. (2.2), the calculated results of sensitivity of each tag are shown in Table 2.2. Among them, the electromagnetic wave in free space wavelength of 0.33 m, the RFID reader the transmission power of P_t is 24 dBm, calculate generation

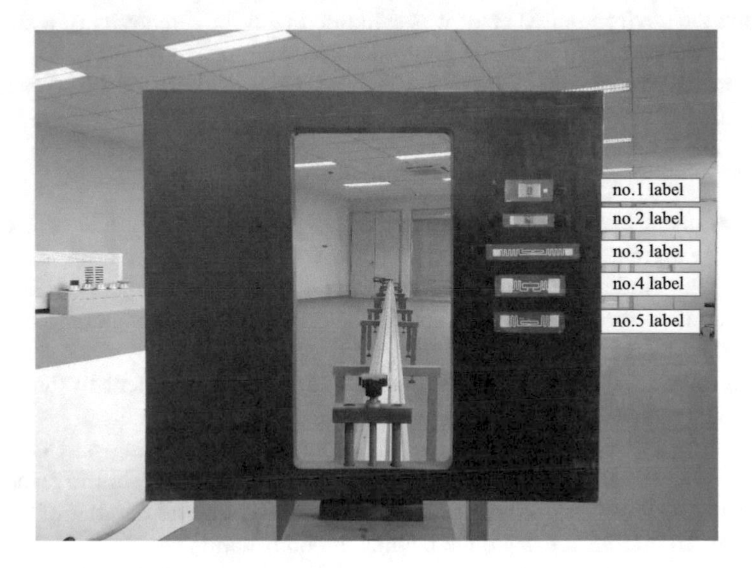

Fig. 2.3 Tag schematic diagram

Table 2.1 Single tag reading distance

Distance/m	No. 1	No. 2	No. 3	No. 4	No. 5
1	4.274	7.015	8.289	11.954	12.092
2	4.132	7.085	8.142	11.926	12.032
3	4.262	7.065	8.282	11.827	12.107
4	4.269	7.124	8.149	11.895	12.087
5	4.102	7.068	8.116	11.914	12.046
Average	4.208	7.071	8.196	11.903	12.073

Table 2.2 Single tag sensitivity

Tag number	Read distance/m	Sensitivity/dBm
1	4.208	−9.111
2	7.071	−13.619
3	8.196	−14.901
4	11.903	−18.143
5	12.073	−18.266

into the value of 251, the RFID reader antenna gain of the G_t is 2 dBi, computing into a value of 1.58, the RFID tag antenna gain G_r 9 dBi, calculate the generation into a value of 7.94, tau antenna matching coefficient is 1. The calculation methods of the input value are shown as follows:

$$G_t : 10\,\mathrm{lg}^{1.58} = 2 \tag{2.5}$$

$$G_r : 10\,\mathrm{lg}^{7.94} = 9 \tag{2.6}$$

$$P_t : 10\,\mathrm{lg}^{251} = 24 \tag{2.7}$$

Table 2.2 shows that the greater the reading distance, the greater the absolute value of sensitivity, that is, the better the tag performance.

(4) Error of calibration

Because it is difficult to measure the environmental disturbance directly, it is necessary to estimate the environmental error indirectly through the instrument. The test environment of single tag and tag group is almost the same, so the environmental error in the experiment is basically the same. The value of tag group sensitivity can be calibrated by subtracting the estimated disturbance value from the value calculated in the experiment. Based on the RFID tag tagformance test equipment, it is often used for performance test of hf and UHF tags, including tag read distance and backscattering power, but the performance of the equipment can only detect a single tag. No obvious interference sources Tagformance test environment, it is generally believed to use Tagformance measurement data is accurate value. Tagformance system structures is shown in Fig. 2.4.

The operating methods of the instrument are as follows:

(1) Connect the system, turn on the power, and turn on the computer supporting software.
(2) Set the frequency range of the test, select the test item (backscattered power or reading distance), place the calibration tag above the instrument in the specified direction, and click the “start” button.

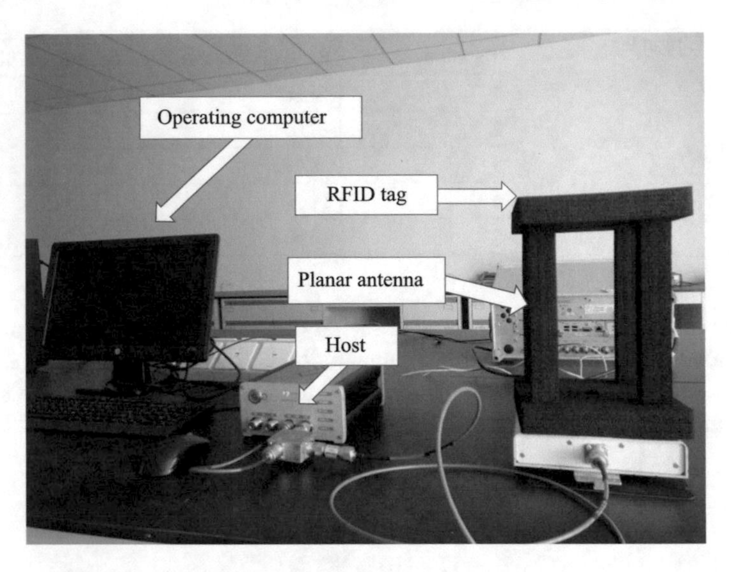

Fig. 2.4 Tagformance system schematic diagram

(3) Observe the error between the test result curve and the reference curve. If the error is within ± 0.5 dBm, it is considered to meet the calibration requirements. Click "clear screen" to prepare for the formal test.
(4) Set the frequency range and step of the test, place the test tag above the instrument according to the polarization direction of the antenna, and click "STARTSWEEP" to start the test.
(5) After the test is completed, save the data and turn off the power.

In this test, the frequency range is set between 800 and 1000 MHz, and the frequency step is 20 MHz. The maximum values of the sensitivity of the five tags in the frequency range were –11.184 dBm, –15.276 dBm, –16.657 dBm, –20.029 dBm, and –20.347 dBm, respectively. The values measured by the instrument are compared with those calculated by the experiment, as shown in Fig. 2.5.

Figure 2.5 shows the comparison results of tag sensitivity obtained by the two methods. Among them, the broken lines of the square mark represent the sensitivity value directly measured by the instrument, the broken lines of the triangle mark represent the sensitivity value calculated according to the reading distance measured by the experiment, the x-coordinate represents the serial number of the tag, and the y-coordinate represents the sensitivity value.

It can be seen that the trends of the two curves in Fig. 2.5 are roughly the same, and the difference between the two curves is the error interference in the experimental environment. Calculate the difference between the two methods of each tag to get the data, and take its average value to get the average error of the tag test environment is 1.891dBm. The error value can be used to correct the test results of tag group sensitivity and eliminate the environmental error to obtain more accurate results.

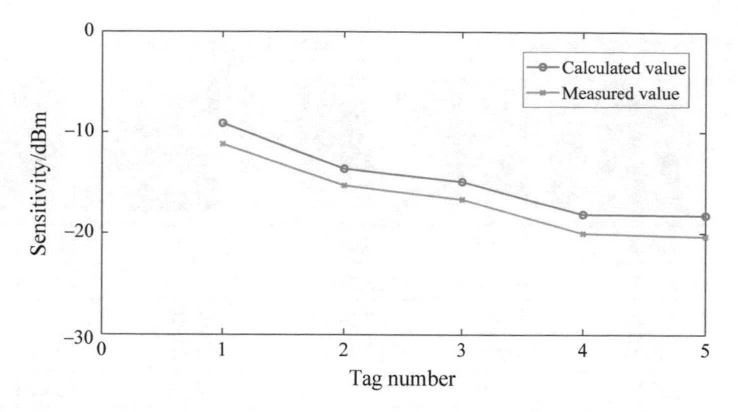

Fig. 2.5 Comparison of results

2.1.4 Tag Group Sensitivity Measurement

(1) The test method

(1) Instrument alignment: a reflector plate is placed on the guide rail, and the laser rangefinder position is adjusted so that the laser beam emitted by the rangefinder can fall on the reflector plate and the beam direction is perpendicular to the reflector plate.

(2) Tag placement: attach the RFID tag used in the experiment to the reflector board, and ensure that the tag and the reflector board fully fit.

(3) Test: click to start the measurement, start the drive motor, the reflector plate starts to move on the guide rail, when the reader detects the tag information, the laser starts to range. When all the tags are 100% read, the test is completed, the result of the reading distance of the tag group is displayed on the display screen, and the CCD completes the collection of the tag image.

(2) The influence of tag structure on tag group sensitivity

In measuring sensitivity tag group at the same time, in order to explore the performance of the worst tag affect the performance of tag clouds, did three groups of experiment, tag number are the same when the position and sequence number unchanged and tag position change, three key tags in the experiment on the baffle position all don't change, and white diamonds to mark the position of key tag. The first is the changes in the reading performance of the tag group when the number of tags is unchanged and the order is changed. When the key tag position changes, the schematic diagram of the tag placement and the measurement results are shown in Table 2.3.

The measurement results of the reading distance of the tag group are shown in Fig. 2.6, where the abscissa represents the five placement methods, and the ordinate

Table 2.3 The recognition distance of tag group when the key tag position changes

Distance/m	Label 1 Label 2 Label 5 Label 4 Label 3	Label 4 Label 1 Label 3 Label 5 Label 2	Label 2 Label 4 Label 1 Label 3 Label 5	Label 5 Label 4 Label 2 Label 1 Label 3	Label 3 Label 5 Label 4 Label 2 Label 1
1	2.413	3.149	3.436	3.070	2.362
2	2.382	3.063	3.394	3.085	2.458
3	2.416	3.121	3.442	3.094	2.423
4	2.396	3.081	3.435	3.131	2.396
5	2.427	3.039	3.383	3.063	2.486
Average	2.401	3.091	3.418	3.089	2.435

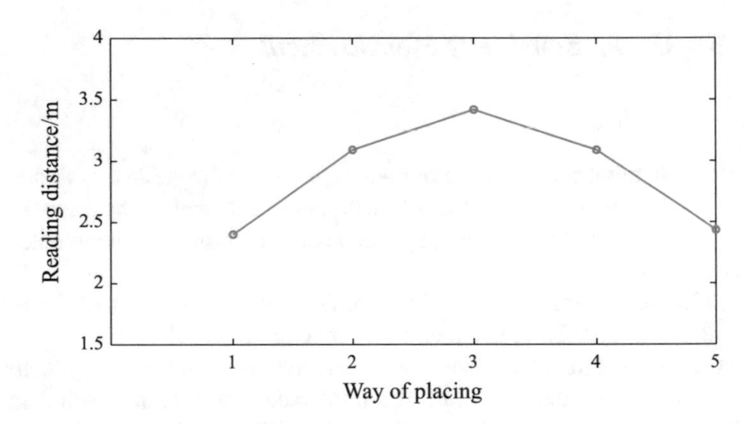

Fig. 2.6 Reading distance when the number of tag groups is unchanged and the structure is changed

represents the average value of the reading distance measurement of each placement method.

Combined with the tag placement diagram and data results, it can be seen that when the key tag is located in the center of the tag group, it has the least impact on the performance of the tag group. When the key tag is located at the edge, it has the greatest impact on the performance of tag group. At this point, the recognition distance of tag group is the smallest, and the absolute value of tag group sensitivity is the smallest. In addition, the maximum reading distance of tag group is less than the reading distance of key tags, that is, the reading distance of tag group is less than the reading distance of a single tag, and the sensitivity value of tag group is less than the sensitivity value of a single tag.

Equation (2.4) is used to calculate the sensitivity of the tag group. Considering that there is no obvious interference source in the laboratory environment, formula (2.4) can be simplified as

$$P_r' = \frac{\lambda^2 P_t' G_t' G_r' \tau'}{(4\pi)^2} \times \frac{1}{r^2} \tag{2.8}$$

Equation (2.8) was used to calculate the sensitivity of tag group of five placement modes. r' is 2.401 m, 3.091 m, 3.418 m, 3.089 m, and 2.425 m, respectively. The wavelength of electromagnetic wave in free space is 0.33 m. The transmission power of the RFID reader is 24dBm, the calculated input value is 251, the gain of the RFID reader antenna is 2dBi, the calculated input value is 1.58, the gain of the RFID tag antenna is 9dBi, the calculated input value is 7.94, and the antenna coefficient is 1. The sensitivity of the five groups was calculated as –4.237 dBm, –6.431 dBm, –7.305 dBm, –6.426 dBm, and –4.324 dBm, respectively. According to the pre-obtained environmental error value of 1.891 dBm, the final values of tag group sensitivity were –6.128dBm, –8.322 dBm, –9.196 dBm, –8.817 dBm, and –6.215 dBm, respectively.

Without changing the experimental environment and experimental Settings, when the number of tags and key tag positions remain unchanged, the schematic diagram and measurement results of tags when the order of other tags changes are shown in Table 2.4. The schematic diagram of the first group of tags is the same as that of the first group in Table 2.3, so the test results in Table 2.3 are directly used.

Equation (2.8) was used to calculate the sensitivity of tag group of five placement modes. r' is 2.401 m, 2.384 m, 2.394 m, 2.393 m, and 2.396 m, respectively. The wavelength of electromagnetic wave in free space is 0.33 m. The transmission power of the RFID reader is 24 dBm, the calculated input value is 251, the gain of the RFID reader antenna is 2dBi, the calculated input value is 1.58, the gain of the RFID tag antenna is 9 dBi, the calculated input value is 7.94, and the antenna coefficient is 1. The sensitivity of the five groups was calculated as –4.237 dBm, –4.176 dBm, –4.212 dBm, –4.208 dBm, and –4.219 dBm, respectively. According to the pre-obtained environmental error value of 1.891 dBm, the final values of tag group sensitivity were –6.128 dBm, –6.067 dBm, –6.103 dBm, –6.099 dBm, and –6.110 dBm, respectively.

Table 2.4 The reading distance of tag group when the seed tag is unchanged

Distance/m	Label 1, Label 2, Label 5, Label 4, Label 3	Label 1, Label 4, Label 5, Label 2, Label 3	Label 1, Label 4, Label 2, Label 5, Label 3	Label 1, Label 4, Label 2, Label 3, Label 5	Label 1, Label 3, Label 5, Label 2, Label 4
1	2.413	2.376	2.369	2.403	2.416
2	2.382	2.407	2.425	2.399	2.379
3	2.416	2.367	2.390	2.376	2.399
4	2.396	2.374	2.376	2.373	2.412
5	2.427	2.396	2.410	2.414	2.373
Average	2.401	2.384	2.394	2.393	2.396

The measurement results of the reading distance of the tag group are shown in Fig. 2.7, where the abscissa represents five placement methods, and the ordinate represents the average reading distance. Combined with the tag placement diagram and data results, it can be seen that the measurement results of the five placement methods are not significantly different, that is, when the position of key tags has been determined, the change of the order position of other tags has little impact on the overall performance of the tag group.

(3) Influence of tag number on tag group sensitivity

Keep the experimental environment and experimental Settings, do not change the tag position, gradually reduce the number of tags. The layout diagram and measurement results are shown in Table 2.5. The layout diagram of the first group of tags is the same as that of the third group in Table 2.3, so the test results in Table 2.3 are directly used. The measurement results of the tag group are shown in Fig. 2.8, where the abscissa represents five placement methods, and the ordinate represents the average reading

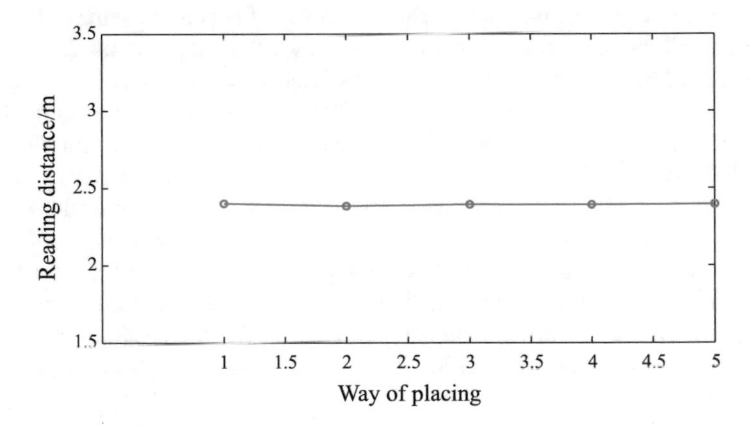

Fig. 2.7 Reading distance when the number of tag groups is unchanged and the order is changed

Table 2.5 The reading distance when the number of multiple tags changes

Distance/m	Label 2 Label 4 Label 1 Label 3 Label 5	Label 4 Label 2 Label 1 Label 3	Label 2 Label 1 Label 3	Label 2 Label 1	Label 1
1	3.436	3.132	3.013	2.589	4.274
2	3.394	3.146	2.995	2.673	4.132
3	3.442	3.148	2.958	2.643	4.262
4	3.435	3.154	3.045	2.654	4.269
5	3.383	3.139	2.957	2.570	4.102
Average	3.418	3.144	2.994	2.626	4.208

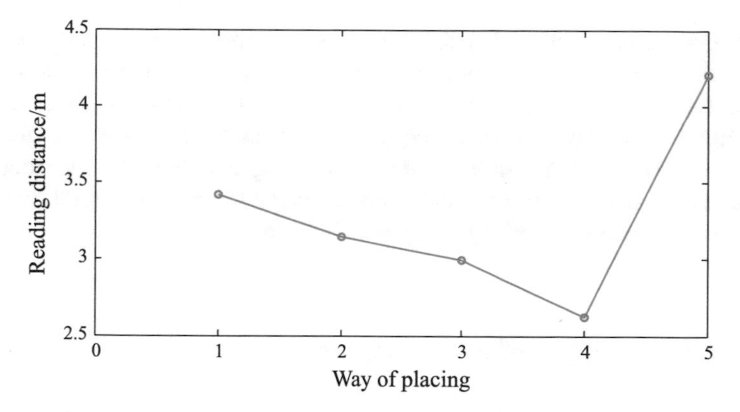

Fig. 2.8 The reading distance when the number of multiple tags changes

distance. Combined with tag put schematic diagram and the data can be seen, tabbed position number changes, the tag of the read distance along with the reduction in the number of tags gradually become smaller, the tag clouds performance gradually along with the reduction in the number of tags, but the tag group of sensitivity value is always less than the value of the sensitivity of key tag, so when the tag number is greater than or equal to two, tag, tag quantity is less, tag clouds the worse performance.

Equation (2.8) was used to calculate the sensitivity of tag group of five placement modes. r' is 3.418 m, 3.144 m, 2.994 m, 2.626 m, and 4.208 m, respectively. The wavelength of electromagnetic wave in free space is 0.33 m. The transmission power of the RFID reader is 24 dBm, the calculated input value is 251, the gain of the RFID reader antenna is 2 dBi, the calculated input value is 1.58, the gain of the RFID tag antenna is 9dBi, the calculated input value is 7.94, and the antenna coefficient is 1. The sensitivity of the five groups was calculated as –7.305 dBm, –6.579 dBm, –6.154 dBm, –5.015 dBm, and –9.111 dBm, respectively. According to the pre-obtained environmental error value of 1.891 dBm, the final values of tag group sensitivity were –9.196 dBm, –8.470 dBm, –8.045 dBm, –6.906 dBm, and –11.002 dBm, respectively.

(4) Influence of tag number on tag group sensitivity

Can be seen from the three test results: tag number unchanged structure changes, the location of the key tags have certain influence the performance of tag clouds, namely when key tags in the group of edge, the tag of the value of the minimum sensitivity, key tags most affected by the performance of tag clouds, on the contrary, when the key tag, in the middle, tag clouds tag group of the value of the maximum sensitivity, key tags to group of minimal performance impact; When the number of multiple tags is unchanged and the order is changed, the position of key tags is unchanged, while other tag changes have little impact on the performance of tag group. Tabbed position unchanged number changes, tag clouds sensitivity value decreased gradually with the reduction in the number of tags, the tag group performance gradually along with the reduction in the number of tags, but the tag group of sensitivity value is always less than the value of the sensitivity of key tag, so when the tag number is greater than or equal to two, tag, tag is less, the number of tag clouds the worse performance.

At the same time, the sensitivity value of tag group is obtained by distance calculation, and the calculated result is calibrated by using the environmental error value, and a more accurate value of tag group sensitivity is obtained.

2.1.5 The Summary of This Chapter

Firstly, we put forward the definition of tag group of sensitivity, and carry on the detailed elaboration, according to the tag of the definition of sensitivity that reverse tag group read distance are needed to determine tag group of the value of the sensitivity to design and set up a dynamic test system based on photoelectric sensor, the specific structure of the system and experimental method are introduced. Then, use the system to complete a single tag read distance test and the calculation of sensitivity, and use the instrument measurement to get the accurate values of the single tag sensitivity, the sensitivity value calculated according to the distance and the difference of the sensitivity of instrument were measured directly to has carried on the quantitative interference existing in the environment, the interference value can be used to calculated value of tag clouds sensitivity calibration. Then, the sensitivity of tag group was tested, and the effects of seed tags on the performance of tag group were investigated under three conditions, namely, the change of structure, the change of sequence, and the change of position of tag group. When the number of multiple tags is unchanged and the structure is changed, that is, when the key tag is located at the edge of the tag group, the key tag has the greatest impact on the performance of the tag group, and when the key tag is located in the middle of the tag group, the key tag has the least impact on the performance of the tag group. When the number of multiple tags is unchanged and the order is changed, the position of key tags is unchanged, while other tag changes have little impact on the performance of tag group. When the number of tags is unchanged, that is, the performance of the tag group gradually becomes worse with the decrease of the number of tags, but the value of the sensitivity of the tag group is always less than the value of the sensitivity of the key tags. Therefore, when the number of tags is greater than or equal to two, the fewer the number of tags in the tag group, the worse the performance of the tag group. Finally, using the designed dynamic test platform to measure the reading distance of each tag group, the sensitivity value of the tag group is calculated, and the calculated result is calibrated with the error value.

2.2 Image Feature Matching Experiment of RFID System Physical Anti-Collision

In the last chapter, we solved the test method of tag group sensitivity and the influence of different tag distribution on tag group sensitivity. When it is necessary to analyze the performance of multiple tag groups, we use the means of image acquisition and processing to confirm the tag position information and complete the evaluation of tag group performance. This chapter will according to the third chapter of RFID tags group dynamic testing system, in view of the different key tag location, image feature matching method to be used for the image obtained by CCD camera system to extract the characteristics of the key tag, key tags in the complex environment of effective feature matching and positioning, image feature matching method to verify the feasibility of performance evaluation of tag clouds. Firstly, SIFT algorithm, SURF algorithm, and ORB algorithm are, respectively, used in this chapter for image matching test. Three algorithms are compared and analyzed, respectively, from the execution time of the algorithm, the number of feature points and the number of correct matching points, and the SURF algorithm which is most suitable for image feature matching is selected. Then, the algorithm is used to locate the key tags in the tag group. This study is of great significance to the evaluation and optimization of RFID multi-tag performance and provides an important basis for upgrading the sensitivity measurement system of tag group to the performance optimization analysis system.

2.2.1 Image Feature Matching Based on SIFT Algorithm

SIFT algorithm can be called scale-invariant feature conversion algorithm, which detects key points in the spatial scale of images and extracts invariants of rotation, scale and position of key points. SIFT algorithm processing speed block, but also has the advantages of scalability and multidimensionality, can solve the image of occlusion, noise, light and other interference on the detection target. The steps of SIFT algorithm can be summarized as generating feature points and describing feature points.

(1) Generate feature points

The detection of the extremum point is to search all the positions on the scale space, and then to detect the extremum point with the scale invariable by differential. First, you need to build a scale space, the SIFT algorithm, using the scale of the image by the Gaussian kernel space, the main reason is that the Gaussian kernel function only measures the same kernel function, can retain more the original image, the characteristics of the Gaussian convolution is expressed as Eqs. (2.9) and (2.10).

Fig. 2.9 Gauss pyramid model

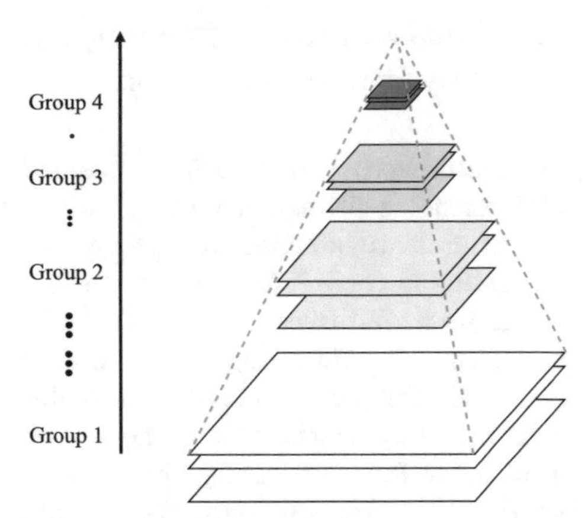

$$G(x_i, y_i, \sigma) = \frac{1}{2\pi\sigma^2}\exp\left[-\frac{(x - x_i)^2 + (y - y_i)^2}{2\sigma^2}\right] \tag{2.9}$$

$$L(x, y, \sigma) = G(x, y, \sigma) \cdot I(x, y) \tag{2.10}$$

where $G(x, y, \sigma)$ represents the Gaussian function, $I(x, y)$ represents the original image, $L(x, y, \sigma)$ represents the convolution operation result of the Gaussian function and the original image, (x, y) represents the pixel position, $*$ represents the convolution operation, and σ represents the fuzzy coefficient.

After obtaining the Gaussian convolution function, it is necessary to establish the Gaussian pyramid to represent the scale space. The pyramid model is shown in Fig. 2.9. The original image is the first group of pyramid models, namely the first layer at the bottom of the pyramid. Each group is composed of six images of the same size but with different fuzzy coefficients. The first layer (the bottom image) of the next group directly samples the penultimate layer of the previous group, thus ensures the continuity of scale space.

After the completion of the Gaussian pyramid, the construction of the pyramid of Difference of Gaussian (DOG) needs to continue. The difference Gauss pyramid is determined by the Gauss pyramid, that is, the second layer of the first group of Gauss pyramid minus the first layer of the first group of Gauss pyramid, the first layer of the first group of Gauss difference pyramid is obtained, and so on. Therefore, Gauss pyramid is the foundation of the pyramid of DOG. In the scale space, the detection of extreme value points is carried out on the basis of the DOG pyramid, that is, every pixel in each layer of the image on the Gaussian difference pyramid is compared with its adjacent 26 pixels, namely, 8 pixels around the same layer and 9 pixels in the upper and lower layers. In this way, the accuracy of detecting extreme value points in scale space can be guaranteed. The detection of extreme value points is shown in Fig. 2.10.

The points obtained above are discrete points and not exact extremum points, so it is necessary to carry out curve fitting for Gaussian difference function. The Gaussian

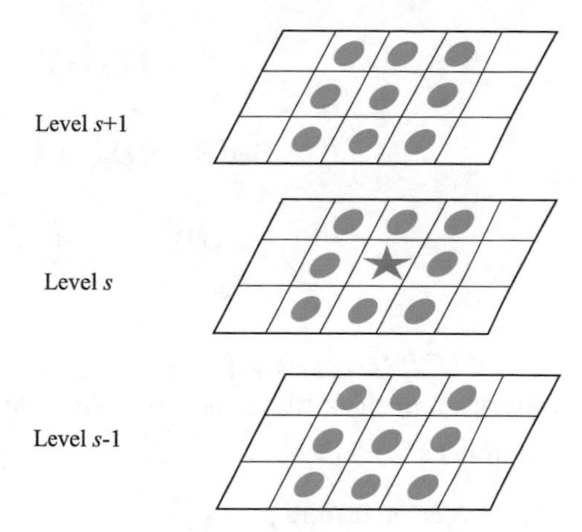

Fig. 2.10 Detection of extremum points

difference function is expressed as Eq. (2.11), the offset of the extreme point $\hat{X}$ can be expressed as Eq. 2.12, and the extreme point $D\left(\hat{X}\right)$ as Eq. 2.13.

$$D(x, y, \sigma) = [G(x, y, k\sigma) - G(x, y, \sigma)] * I(x, y) = L(x, y, k\sigma) - L(x, y, \sigma) \tag{2.11}$$

$$\hat{X} = -\frac{\partial^2 D^{-1}}{\partial X^2}\frac{\partial D}{\partial X} \tag{2.12}$$

$$D\left(\hat{X}\right) = D + \frac{1}{2}\frac{\partial D^{\mathrm{T}}}{\partial X}X \tag{2.13}$$

where $X = (x, y, \sigma)$ represents the coordinates of the spatial extreme point, and k represents the number of layers, leaving only extreme points with a contrast greater than 0.04.

Remove the low contrast points and eliminate the edge effect, so as to get more stable extremum points. The principal curvature of the Gaussian difference function is smaller at the vertical edge of the extremum and larger at the point across the edge. The main curvature of Gauss difference function represented by matrix is

$$H = \begin{bmatrix} D_{xx} & D_{xy} \\ D_{xy} & D_{yy} \end{bmatrix} \tag{2.14}$$

where D_{xx}, D_{xy} and D_{yy} are, respectively, obtained by neighborhood difference of each point. Let the maximum eigenvalue of H be γ, and the minimum eigenvalue of H be ε. Only the ratio between D and H is considered, then

$$\gamma = r\varepsilon \tag{2.15}$$

$$T_r(H) = D_{xx} + D_{yy} \tag{2.16}$$

$$\mathrm{Det}(H) = D_{xx} \times D_{yy} - D_{xy} \times D_{xy} \tag{2.17}$$

$$\frac{T_r(H)^2}{D_{et}(H)} = \frac{(\gamma + \varepsilon)^2}{\gamma\varepsilon} = \frac{(\gamma + 1)^2}{\gamma} \tag{2.18}$$

${(r+1)^2}/{r}$ increases with the increase of r, and the minimum value is obtained when the two eigenvalues are equal. This point is retained when the ratio of the two eigenvalues is less than ${(r+1)^2}/{r}$.

(2) Describe feature points

After the accurate extremum point is obtained, the method of image gradient is needed to obtain the reference direction of the local structure of the key points, that is, the gradient amplitude and gradient direction of each key point are calculated, so that the descriptors have invariance to the image transformation. The gradient amplitude and gradient direction of the key points were calculated by histogram, as shown in Fig. 2.11. The histogram ranges from 0° to 360° and is divided into 36 columns on average. The peak value of the histogram is the main direction of the key point, representing the direction of the neighborhood gradient of the point, and the auxiliary direction is the direction greater than 80% of the main direction in the histogram. The histogram was smoothed and the Taylor expansion was fitted by quadratic method to get the exact direction of the key points.

At this point, the position, scale and direction of the key points of the image are determined, that is, it has the properties of translation, scaling and rotation invariability, and the next step is to describe the contribution points of the key points and

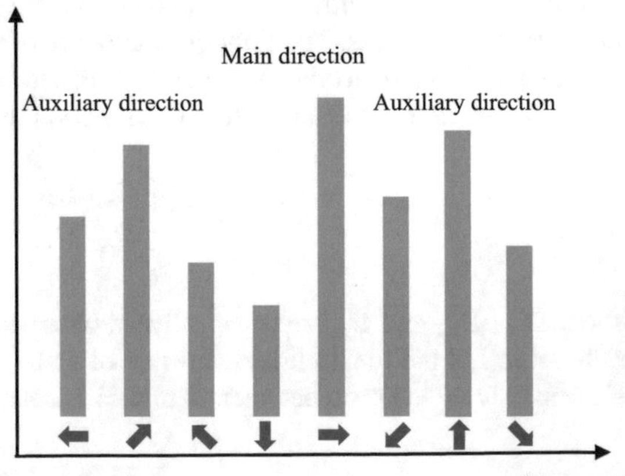

Fig. 2.11 Histogram of key points

their surroundings. It is represented by a vector descriptor of $4 \times 4 \times 8 = 128$ dimensions. Firstly, the image region required by the descriptor is determined. Secondly, change the coordinate Angle and take the main direction of the key point as the X-axis direction; then, the gradient amplitude and gradient direction of each pixel in the image region are calculated, and the histogram of pixel is generated. Finally, the vector elements are normalized to limit the gradient amplitude on the histogram below a threshold.

After you have completed the above steps, you can make a key point match between the reference graph and the observation graph. In general, the key point matching is mainly based on the key point feature of the reference image.

2.2.2 Image Feature Matching Based on SURF Algorithm

SURF algorithm is fast and robust. SURF algorithm can well deal with the matching problem between the reference image and the observation image in different situations, such as rotation, zoom, and brightness change. Actually, SURF algorithm is based on SIFT algorithm using the Hessian matrix to improve the speed, robustness enhancement mainly because using the Haar wavelet transform, the algorithm can be used for 3 d reconstruction in the field of machine vision, etc. The realization process of SURF algorithm mainly consists of feature point generation and feature point description.

(1) Generate feature points

Hessian matrix is a matrix composed of the second partial derivatives of a multivariate function, and is also an important part of SURF algorithm. The local maximum value of the determinant of the matrix is the position of the key point. The box filter is a tool to solve the arithmetic of adding and subtracting pixels in the neighborhood. The determinant of Hessian matrix is calculated by using the box filter. The Hessian matrix can be expressed as (2.19), and the determinant of the matrix is (2.20)

$$H(x, \psi) = \begin{bmatrix} L_{xx}(x, \psi) & L_{xy}(x, \psi) \\ L_{xy}(x, \psi) & L_{yy}(x, \psi) \end{bmatrix} \tag{2.19}$$

$$\begin{aligned} \mathrm{Det}(H) &= D_{xx} \times D_{yy} - D_{xy} \times D_{xy} \\ &= D_{xx} \times D_{yy} - \left(\rho * D_{xy}\right)^2 \end{aligned} \tag{2.20}$$

where x represents the pixel position, ψ represents the scale, ρ is used to balance the error of using the box filter, and $L_{xx} L_{xy} L_{yy}$ represents the second derivative of the image in all directions after Gaussian filtering.

In the SURF scale space, the size of the inter-group images is the same, but the size of the box filter is different. The response image of the determinant of Hessian

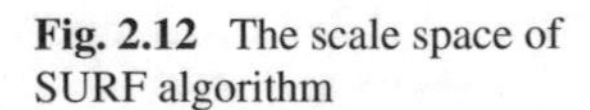

Fig. 2.12 The scale space of SURF algorithm

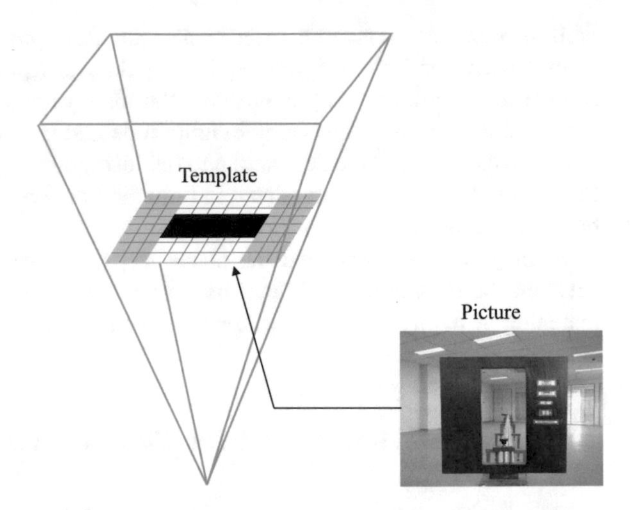

matrix is obtained by using the integral image of the original graph, and the extremum points of different scales are obtained. The scale space of SURF algorithm can be shown in Fig. 2.12.

Similar to SIFT algorithm, each pixel point is compared with 26 points in its neighborhood, leaving a maximum point. If the maximum point is less than the threshold of the Hessian determinant, this point will be excluded.

(2) Describe feature point

After finding the feature points, the next step is to determine the direction of the feature points. In SURF algorithm, feature points are taken as the center of a circle and a circular neighborhood is determined. A 60-degree sector area is selected every 0.2 radians, and all points of small porter's feature are counted. The direction with the most points is the main direction of the feature point, as shown in Fig. 2.13. After determining the main direction of feature points, the image is divided into 4×4 blocks along the main direction, and then the Haar wavelet template is used to calculate each block, and the statistical calculation result is the feature vector of feature points.

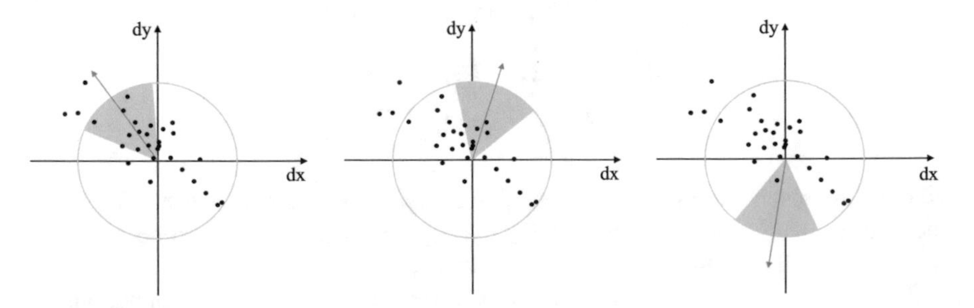

Fig. 2.13 Determine the main direction

2.2.3 Image Feature Matching Based on ORB Algorithm

ORB algorithm is a feature matching algorithm that combines FAST algorithm and BRIEF algorithm, two high-performance and low-workload methods. The biggest advantage of FAST algorithm is its FAST speed, and BRIEF algorithm is used to describe the direction of feature descriptors. Because of the speedup of the FAST algorithm, the ORB algorithm runs much faster than SIFT and SURF algorithms, and the descriptors obtained by using the BRIEF algorithm can effectively save storage space. The scale variability of the ORB algorithm is not as good as the other two algorithms. The principles of the ORB algorithm can also be summarized by generating and describing feature points.

(1) Generate feature point

FAST algorithm is to compare the gray difference between the selected pixel point and the area around the point. If the difference is greater than a certain threshold, the point can be used as a feature point. The implementation process of FAST algorithm consists of three steps:

(1) By detecting all the pixels on the circle with a certain radius, the unqualified pixels can be preliminarily removed and possible corner points can be left. Take fast-x–y as an example (X, Y as a number), select Y pixels on the edge of the circle and make a difference with the grayscale of the pixel at the center of the circle. If there are continuous X points that satisfy the condition, that is, the difference between the grayscale of these points and the grayscale of the center pixel is greater than a certain threshold, then this point can be selected as a feature point. The formula used can be described as

$$S_{p\to x} = \begin{cases} b, \ I_{p\to x} \le I_p - \varepsilon & \text{lightless} \\ u, \ I_p - \varepsilon < I_{p\to x} < I_p + \varepsilon & \text{similar} \\ g, \ I_p + \varepsilon \le I_{p\to x} & \text{partial light} \end{cases} \tag{2.21}$$

where I_p represents the pixel value at the center of the circle, $I_{p\to x}$ represents the pixel value at the point x on the circle, and ε is the threshold value. Therefore, the meaning of Eq. (2.21) is when $I_{p\to x} \le I_p - \varepsilon$, that is, when the pixel value at the center of the circle is greater than the pixel value at x, the point is dark; when $I_p - \varepsilon < I_{p\to x} < I_p + \varepsilon$, that is, the pixel value at the center of the circle is similar to the pixel value at x. When $I_p + \varepsilon \le I_{p\to x}$, that is, the pixel value at the center of the circle is less than the pixel value at x, the point is brighter. According to this equation, the circular region can be divided into three parts: b, u, and g. As long as the number of occurrences of b and u is counted, the candidate corner points can be easily determined.

(2) The classifier is used to detect the selected feature points and judge whether they satisfy the corner feature. The pixels divided into three parts can be represented as P_b, P_u, and P_g, respectively. Define a variable K_p to mark the information of

the classification. If p is the corner, it is the real information. Variable K_p only collects the real information. The gain of pixel is determined by the entropy value of variable K_p, and the maximum gain is used as the judgment condition, that is, whether the point is an Angle point or not.

(3) After completing the first two steps, the resulting corner points need to be validated. Firstly, the response function V of the diagonal points needs to be defined, that is, the intensity of the corner points is the sum of the pixel values of several corner points on the center of the circle and the edge of the circular template. The function V can be expressed as

$$V = \max\left(\sum\nolimits_{x \in S_{\text{lighter}}} \left|I_{p\to x} - I_p\right| - \varepsilon, \sum\nolimits_{x \in S_{lightless}} \left|I_p - I_{p\to x}\right| - \varepsilon\right) \tag{2.22}$$

The response function of adjacent two points is calculated, the local search is carried out, the non-maximum value is suppressed, and the corner point, namely the characteristic point, is obtained.

(2) Describe feature points

The FAST algorithm mainly solves the problem of finding key points in the image, and then the BRIEF algorithm is needed to turn these feature points into feature vectors. The operator of BRIEF is composed of binary bits, expressed as

$$Z(p, x, y) = \begin{cases} 1 \; p(x) < p(y) \\ 0 \; p(x) \geq p(y) \end{cases} \tag{2.23}$$

$p(x)$ and $p(y)$, respectively, represent the gray value at x and y points. That is, if the gray value at x is greater than or equal to the gray value at y, the corresponding binary value is 0; if the gray value at y is less than, the binary value is 1. The feature description operator is shown in Eq. (3.16)

$$f_n(p) = \sum_{1 \leq i \leq n} 2^{i-1} Z(p_i, x_i, y_i) \tag{2.24}$$

All pixels were tested using Eq. (3.15) and the average value $\overline{Z}$ was calculated. You compare the difference between Z and $\overline{Z}$, and you sort it to get a vector Z. Remove the points with large differences from Z and place them in a new vector K. After all the points in the vector Z are screened, the rest of the points need to be tested for their correlation with the vector K, until 256 pairs are satisfied in the vector K, and the operation ends.

2.2.4 Algorithm Comparison Results and Analysis

The three matching methods were tested by arranging the images with tags. The feature points of the reference images were set to 200, and the feature points of the images to be registered were set to 200. This experiment was completed on MATLAB 2013a software in Windows 10 system. The test results are shown in Figs. 2.14, 2.15 and 2.16.

The three algorithms are compared and analyzed, respectively, from three aspects of execution time, the number of feature points, and the number of correct matching points. The results are shown in Table 2.6.

From the above results, it can be seen that for images with certain deviation, the three algorithms can show good matching results. From the point of view of matching time, the execution speed of ORB algorithm is the most advantageous, but the gap between the three algorithms is not big. In terms of the success rate of pairing, the success rate of SIFT algorithm is 72.66%, the success rate of ORB algorithm is 65.46%, and the success rate of SURF algorithm is 81.05%, which is the highest. Since this paper studies the tag position and requires a high matching success rate, this paper selects SURF algorithm to complete image feature matching.

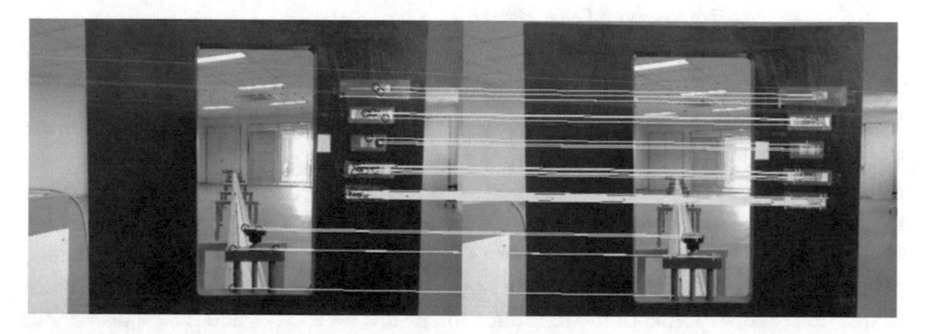

Fig. 2.14 SIFT algorithm results

Fig. 2.15 SURF algorithm results

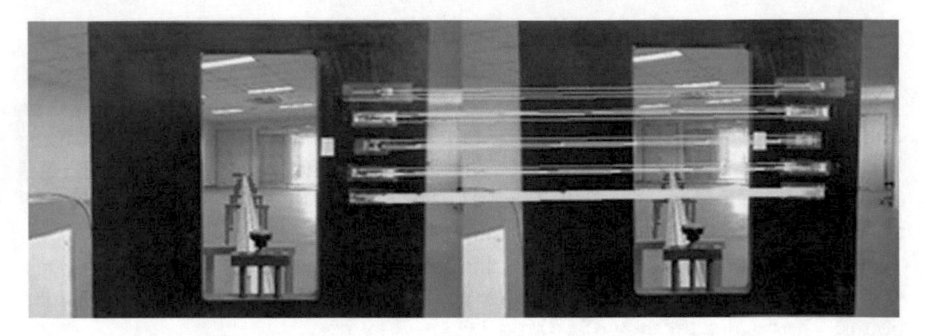

Fig. 2.16 ORB algorithm results

Table 2.6 Test results of the three algorithms

	SIFT	SURF	ORB
Match time/s	1.273	1.047	0.566
Feature points/per unit	168	183	244
Successful pairing	133	167	147

2.2.5 *Image Feature Matching*

The five tags selected in the image matching experiment are shown in Fig. 2.17. The reading distance of five single tags is tested, respectively. The test results are shown in Table 2.7. Among them, the tag with the worst performance is the key tag of this group, and its position is marked with white block. The key tag is used as the matching image of this experiment, as shown in Fig. 2.18.

According to the realization steps of SURF algorithm in Sect. 2.3.2, the realization process of SURF algorithm is shown in Fig. 2.19. Duc to the small size of the reference image, the feature points of the reference image are set to 100, and the feature points of the image to be registered are set to 300. Where (a) shows the feature point found on

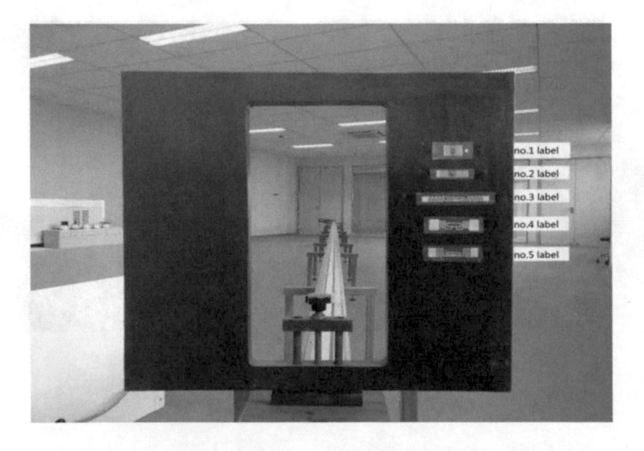

Fig. 2.17 Tag schematic diagram

Table 2.7 Feature matching tag reading distance

Distance/m	1	2	3	4	5
1	4.196	7.086	8.645	11.123	11.851
2	4.232	7.113	9.104	11.083	11.882
3	4.214	7.091	8.726	10.801	11.906
4	4.199	7.098	9.067	10.843	11.879
5	4.182	7.128	9.108	11.303	11.914
Average	4.205	7.103	8.930	11.031	11.886

Fig. 2.18 Matching image

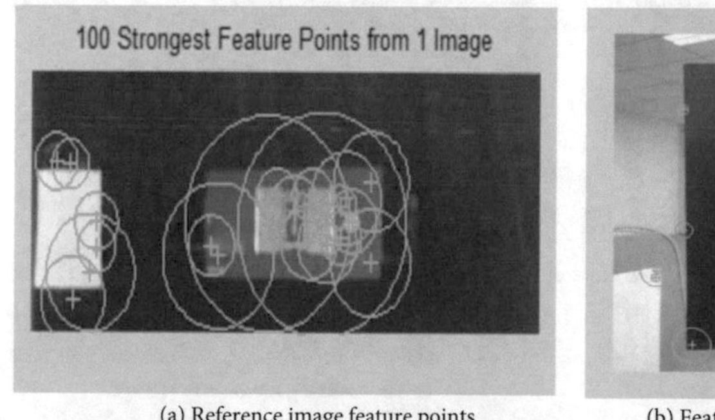

(a) Reference image feature points

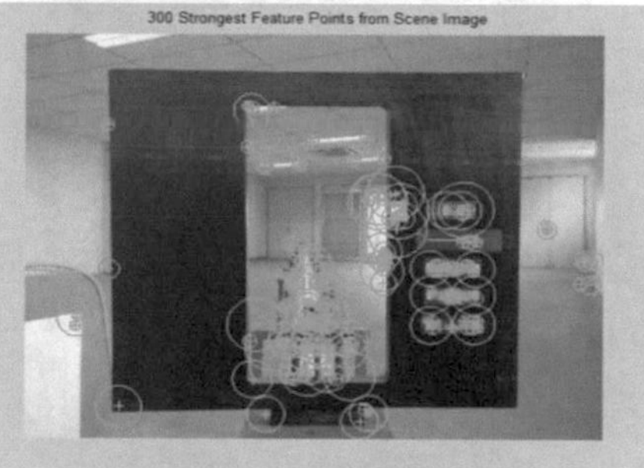

(b) Feature points of the image to be registered

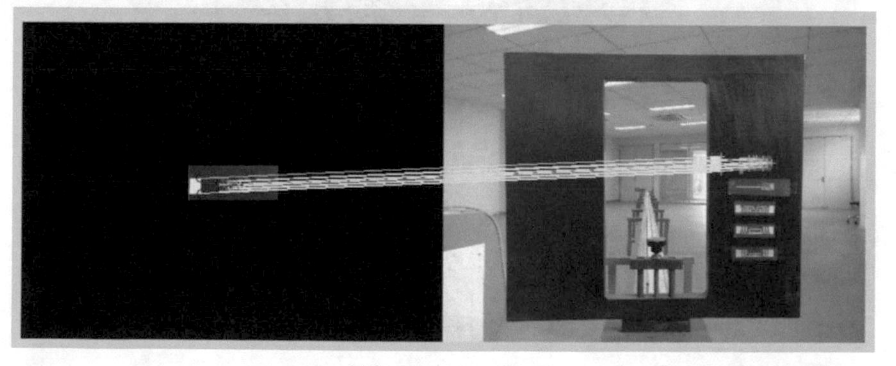

(c) Feature point matching

Fig. 2.19 Implementation steps of SURF algorithm

the reference image, (b) shows the feature point found on the image to be registered, and (c) shows the corresponding feature point of the two images.

The position of the seed tag in the tag group was changed, respectively, and Fig. 2.19 was used as the reference image, which was matched with the tag arrangement diagram of the five positions, respectively, and the results were shown in Fig. 2.20a–e are matched and displayed according to the characteristics of the five

(a) The first tag matches the image

(b) The second tag matches the image

(c) The third tag matches the image

(d) The forth tag matches the image

(e) The fifth tag matches the image

Fig. 2.20 Matching results of seed tags in tag group

groups of tag arrangement diagrams according to the reference image, and the tag position was marked with a rectangular box.

According to the experimental results, SURF algorithm can effectively match the position of key tags in the tag group. When batch identification of tag groups is carried out, SURF algorithm is used to carry out feature matching of the collected tag distribution images. Combined with the conclusion of the influence of key tags on the sensitivity of tag groups obtained in this chapter, the performance of tag groups can be evaluated and analyzed according to the results of image processing.

Combining photoelectric sensing technology and image processing technology, tags distribution image is processed and analyzed based on the data by RFID tag dynamic test system. From the execution time, feature point matching point number and the correct number of SIFT algorithm, SURF algorithm and compares and analyzes the ORB algorithm and choose SURF algorithm according to the results of the comparison of multiple sets of tag distribution characteristic of the image matching, the experimental results show that SURF algorithm can accomplish the key tag positioning, so as to realize performance analysis of the RFID tag group. This chapter is of great significance to solve the evaluation and optimization analysis of RFID multi-tag performance, at the same time, it provides an important research foundation for upgrading the tag group sensitivity measurement system to the performance optimization analysis system.

2.3 Conclusion

Because RFID tags have the advantages of long reading distance, high transmission efficiency, and wide range of functions, they are used in logistics, intelligent transportation systems, and personnel positioning. As the application range of RFID tags becomes more and more extensive, the working environment becomes more and more complex, and the collision problem of multiple tags cannot be ignored. Therefore, higher requirements are placed on the stability of the performance of RFID multiple tags. This research mainly focuses on the performance evaluation of batch identification of RFID tag groups, proposes a test and calibration method for tag group sensitivity, and builds a dynamic test platform for tags based on photoelectric sensing. The feasibility of the proposed tag group sensitivity test and calibration method was verified through experiments. At the same time, according to the research results of the tag sensitivity in the tag group due to the different placement of the tag, the performance of the tag group batch recognition was combined with image feature matching evaluation provides a new way. This study is of great significance to the performance evaluation of tag groups and tag groups in batch identification.

References

1. Liu XL, Xie X, Wang K et al (2017) Pinpointing anomaly RFID tags: situation and opportunities. IEEE Netw 31(6):40–47
2. Anum SJ, Habib A, Anam H et al (2018) Miniaturized humidity and temperature sensing RFID enabled tags. Int J RF Microwave Comput-Aided Eng e21151
3. Wang B, Zhuang YQ, Li XM et al (2016) Design of a novel dual ports antenna to enhance sensitivity of handheld RFID reader. Int J Microw Wirel Technol 8(2):369–377
4. Petrariu AI, Popa V (2016) Analysis and design of a long range PTFE substrate UHF RFID tag for cargo container identification. J Electr Eng-Elektrotechnicky Casopis 67(1): 42–47
5. Boaventura A, Carvalho NB (2016) Measurement of sensitivity improvement in RFID tags. In: IEEE MTT-S international microwave symposium (IMS). San Francisco, CA, pp 1–3.
6. Beuster N, Ihlow A, Blau K (2019) A versatile, automated, cost-effective testing platform for hands-on UHF RFID measurements. In: 2019 IEEE international conference on RFID. Phoenix AZ, pp 1–7
7. Moutis ML, Ennasar MA, Aznabet I et al (2018) A low cost automated RFID tag antenna measurement set-up based on UHF-RFID reader. In: 2018 6th international conference on multimedia computing and systems (ICMCS)
8. Yuan L (2019) Research and practice of RFID-based warehouse logistics management system. In: 2019 international conference on smart grid and electrical automation. Xiangtan, pp 514–519

Chapter 3
Physical Theory of RFID System Physical Anti-Collision

Radio-frequency identification (RFID) technology is a wireless communication technology that enables users to uniquely identify tagged objects or people. An RFID tag is mainly composed of antenna, chip, and Integrated Circuit (IC). The reliability of the RFID tag refers to be read completely under prescribed conditions. Once exceeding the normal environmental conditions (temperature, humidity, and so on), RFID tags could be read difficultly or have poor reading performance because of drifted threshold voltage, long delay time, and inferior noise tolerance [1, 2].

In practical applications, radio-frequency identification (RFID) systems are susceptible to various factors (metals, liquids, temperature, etc.). This chapter takes the reading distance of the tag as the evaluation standard of the reading performance of the RFID system, focusing on the influence of different factors on the dynamic performance of the UHF RFID system. First, an experimental system includes a temperature control system and a detection platform designed to control the temperature and measure the reading distance of the RFID tag, and test the effect of different temperatures on the performance of the tag. Secondly, a theoretical formula for calculating the reading distance of RFID in salt fog and humidity environments is proposed. An RFID dynamic identification experiment platform based on salt fog and humidity environment was designed and established. The relationship between salt fog, humidity, and RFID reading distance was established by fitting experimental data. Finally, a test method for performance evaluation and optimization of radio frequency identification and multiple input multiple output (RFID-MIMO) systems based on Cramer–Rao boundary (CRB) is studied. The effect of the distribution of multiple tags in the RFID-MIMO system on reading performance is analyzed.

X. Yu et al., *Physical Anti-Collision in RFID Systems*,
https://doi.org/10.1007/978-981-16-0835-3_3

3.1 Thermodynamic Analysis of Physical Anti-Collision—Research on the Effect of Temperature on Tag Performance

The study of temperature on the antenna has been reported. In [3], Kabacik presented a thorough study into the performance of microstrip patch antennas that were exposed to large temperature variations and pointed out that the electrical characteristics of microstrip antennas were considerably influenced by temperature. Yadavused used HFSS to study the effect of temperature on microstrip antenna and the results showed that the bandwidth of antenna remained unchanged and the impedance increased with the increase of temperature [4]. Cheng designed a temperature sensor due to the antenna's sensitization to temperature and found that the resonant frequency of the antenna decreased with the increase of temperature [5]. Babu found that high temperatures can cause degradation of the antenna's materials and it would not function well if the ground plane began to fall apart [6].

However, the above researches are relevant to the effect of temperature on antennas of chip and IC in the ideal case. Goodrum discovered that low temperature could lead to difficulty in detecting active tags with a short-read range. When the temperature was as low as −10 °C, RSSI readings were much lower than tags of 22 °C and represented poor RFID performance [7]. Merilampi analyzed the effects of temperature on printed passive UHF RFID tags on a paper substrate and found that the temperature had a severe effect on the tag performance [8]. But the above experiments are about the static tests on the antenna in a closed space with little reference to the dynamic tests.

The main theme of the section is to study the influence of temperature on the RFID tag's dynamical reading performance. We derived an identical equation on reading distance and working frequency of RFID tag, and then converted it to the one on reading distance and environment temperature. Subsequently, the experimentation system, including the temperature control system and experimentation platform, was designed. According to the result of measurement, a fitting model between temperature and reading distance was established and the threshold temperature of UHF tags was obtained. Finally, a temperature compensating mechanism was derived to get the reading distance of a tag at the same reference temperature. The numerical and actual experiments showed that this method was applied to the dynamic measurement of RFID tag's reading performance, which had the advantages of high measuring accuracy, stable performance, and quick response speed.

3.1.1 *Fundamental Principles*

(1) Theory of heat transfer

The environment temperature affects the performance of tags via heat conduction, heat convection, and heat radiation.

(1) Heat conduction: The basic law of heat conduction is given by [9]

$$\phi = -kA\frac{\partial t}{\partial n} \tag{3.1}$$

where ϕ is heat flux, k is heat conductivity coefficient and $\frac{\partial t}{\partial n}$ is normal temperature gradient.

From Eq. (3.1), the direction of the temperature gradient is opposite to the one of heat flow. Thermal conductivity presents the ability of a material's heat conduction and the main factors influencing the thermal conductivity are the kinds of material, temperature, and so on.

(2) Heat convection is defined by the Newton cooling law [10]:

$$\phi{=}Ah_c(t_w - t_f) \tag{3.2}$$

where h_c is the coefficient of convective heat transfer, which presents the transferred heat for 1 °C on per area, A is the area of a solid's surface, t_w is the temperature of fluid, and t_f is the temperature of the solid's surface.

(3) Heat radiation is given by [11]

$$\frac{\Phi_A}{\Phi_0} + \frac{\Phi_R}{\Phi_0} + \frac{\Phi_D}{\Phi_0} - \alpha{+}\beta + \gamma = 1 \tag{3.3}$$

where Φ_0 is the radiation power fall on the tag, Φ_A is the absorbed one, Φ_R is the reflected one, Φ_D is the penetrated one, α is absorptivity, β is reflectivity and γ is transmittance.

The value of absorptivity, reflectivity, and transmittance are related to the tag's nature and will change with environment temperature and the tag's radiation wavelength. For most materials, thermal radiation is not easy to penetrate:

$$\alpha{+}\beta{=}1 \tag{3.4}$$

(2) Reading distance of RFID system

In practical application, RFID tags are attached to the surface of targets and generate induction current to send data due to the electromagnetic field transmitted by reader antennas. Then, the readers detect and decode the backscatter signal of tags. Eventually, the readers send the data of tags to the background processor and the RFID system achieves the purpose of automatic identification of goods.

Reading range is an important characteristic parameter of passive RFID tags. The power density of an electromagnetic wave incident on the RFID-tag antenna in free space is given by [12]

$$S = \frac{P_{tx}G_{tx}}{4\pi R^2} = \frac{P_{EIR}}{4\pi R^2} \tag{3.5}$$

where P_{tx} is the transmitted power, G_{tx} is the gain of the reader's transmitting antenna, R is the distance to the tag, and P_{EIR} is the effective radiated power of the transmitting antenna.

The power P_{tag}, collected by the tag antenna, is by definition the maximum power that can be delivered to the complex conjugate matched load:

$$\begin{aligned} & P_{\text{tag}} = A_e S = \tfrac{\lambda^2}{4\pi} G_{\text{tag}} S = P_{tx}G_{tx}G_{\text{tag}}\left(\tfrac{\lambda}{4\pi R}\right)^2 = P_{tx}G_{tx}G_{\text{tag}}\left(\tfrac{c}{4\pi Rf}\right)^2 \\ & \alpha+\beta = 1 \\ & \phi = Ah_c(t_w - t_f) \\ & \tfrac{\Phi_A}{\Phi_0} + \tfrac{\Phi_R}{\Phi_0} + \tfrac{\Phi_D}{\Phi_0} = \alpha+\beta+\gamma = 1 \end{aligned} \tag{3.6}$$

where λ is the working wavelength between the tag and reader, c is the speed of light, f is the working frequency between the tag and reader, G_{tag} is the tag antenna's gain, and A_e is the effective area of the tag's antenna and is given by

$$A_e = \frac{c^2}{4\pi f^2} G_{\text{tag}} \tag{3.7}$$

The backscatter power from the tag is expressed by [11]

$$P_{\text{back}} = S\sigma = \frac{P_{tx}G_{tx}}{4\pi R^2}\sigma = \frac{P_{EIR}}{4\pi R^2}\sigma \tag{3.8}$$

where σ is the radar cross section of the RFID tag. The backscatter power density from the tag is given by

$$S_{\text{back}} = \frac{P_{tx}G_{tx}}{(4\pi)^2 R^4}\sigma \tag{3.9}$$

The power received by the receiving antenna of the reader can be calculated from the classical radar equation as

$$P_{rx} = A_W S_{\text{back}} = \frac{P_{tx}G_{tx}G_{rx}c^2}{(4\pi)^3 f^2 R^4}\sigma \tag{3.10}$$

where G_{tx} is the gain of the reader's receiving antenna and A_W is the effective area of the reader's antenna which is given by

$$A_W = \frac{c^2}{4\pi f^2} G_{rx} \tag{3.11}$$

The reading distance of passive backscatter RFID system is given by

$$R = \left(\frac{P_{tx} G_{tx} G_{rx} c^2 \sigma}{(4\pi)^3 f^2 P_{rx}} \right)^{1/4} \tag{3.12}$$

(3) Temperature effect on reading distance

In the RFID system, the backscattering signals of passive tags have the same carrier frequency as the transmitting signals of reader, which is the frequency of the reader that is used to activate RFID tags. The temperature change has an effect on the center frequency of the passive tags' backscattering signals. Moreover, the frequency shift is a result of temperature variation. Let us assume that the frequency reduced gradually with the increase of temperature and has a linear relationship with temperature.

Therefore, due to the offset of backscattering signals' center frequency caused by temperature variation, Eq. (3.12) could be rewritten as

$$R = \left(\frac{P'_{tx} G'_{tx} G_{rx} c^2 \sigma}{(4\pi)^3 P_{rx} f'^2} \right)^{1/4} = \left(\frac{P'_{tx} G'_{tx} G_{rx} c^2 \sigma}{(4\pi)^3 P_{rx} (aT + b)^2} \right)^{1/4} \quad (f_1 \leq f' \leq f_2) \tag{3.13}$$

where R is the distance where the signals of RFID tag could be read by a reader, P'_{tr} is the power of backscattering signals, G'_{tx} is the gain of tag's antenna, f' is the migrated frequency of backscattering signals' center frequency, T is environment temperature around RFID tag, and a, b are undetermined coefficients. f_1 and f_2 are the lower frequency limit and upper frequency limit. RFID reader and tag could communicate under the frequency range which is determined by the characteristics of reader. When the frequency f' is less than f_1 or more than f_2, the tag could not be read. When the parameters of a tag are determinate and the tag could be read once the value of signal strength reaches the activated threshold, Eq. (3.13) could be rewritten as

$$R = \eta \left(\frac{1}{(aT + b)^2} \right)^{1/4} \tag{3.14}$$

$$\eta = \left(\frac{P_{tx} G_{tx} G_{rx} c^2 \sigma}{(4\pi)^3 P_{rx}} \right)^{1/4} \tag{3.15}$$

According to Eq. (3.15), η changes with the received power P_{rx}, which is related to the distance between the reader antenna and tag. The focus of this work is the influence of temperature on the dynamic reading performance of the UHF RFID system, so the received power P_{rx} is supposed to be the sensitivity of the reader's antenna. In this case, the received power P_{rx} is constant and then η could be considered as a constant in the simulation.

Since a, b, η are constants, Eq. (3.14) can be rewritten as

$$R = (cT + d)^{-1/2} \text{ or } M = cT + d \tag{3.16}$$

where $c = a/\eta^2$, $d = b/\eta^2$, relevant reading distance $M = 1/R^2$.

$$\sigma = 0.036/\text{m}^2, P_{rx} = -70 \text{ dBm}, a = 914.87$$

Let $P_{tx} = 30\text{dBm}$, $G_{tx} = G_{rx} = 8\text{dBi}$, $\begin{array}{l} \frac{R_1}{R_2} = \left(\frac{cT_2+d}{cT_1+d}\right)^{1/2} \\ T = S - L \\ R_i = (T^2 + H_i^2)^{1/2} \end{array}$, we could get the relationship between the tag's reading distance and temperature in static state, shown in Fig. 3.1.

Realistic test is done in several different temperatures, so the corresponding coefficients and temperature compensation could be obtained. When the corresponding coefficients$^{c,\,d}$ are obtained, the mechanism could be given by

$$\frac{R_1}{R_2} = \left(\frac{cT_2 + d}{cT_1 + d}\right)^{1/2} \tag{3.17}$$

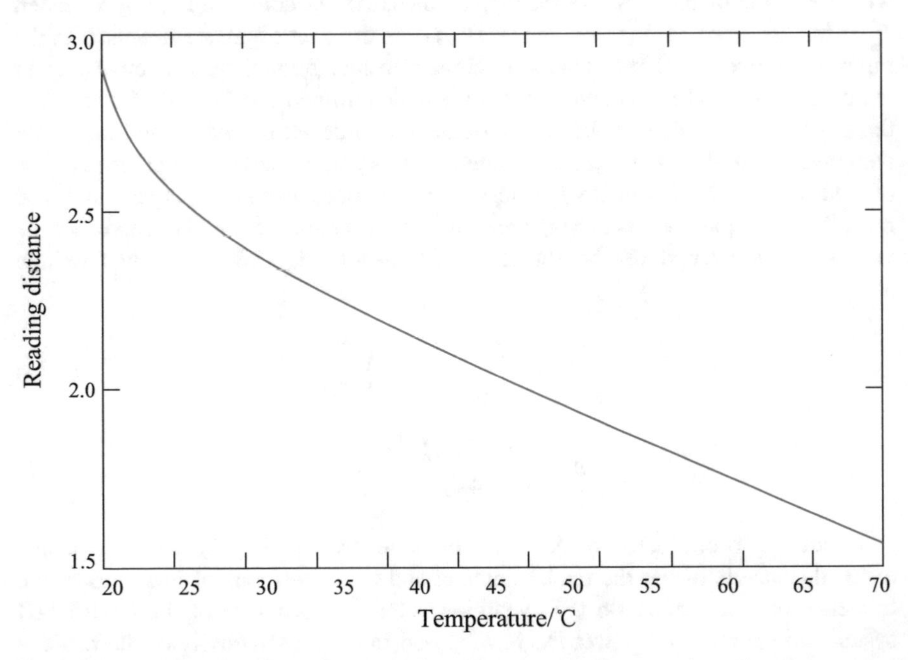

Fig. 3.1 Relationship between the tag's reading distance and temperature in static state

3.1.2 Design of the Experimentation System

(1) Temperature control system

The temperature control system is designed as Fig. 3.2, in which a temperature controller is adapted to measure and control surrounding temperature [13–15]. The system is devised to simulate the influence of temperature on the dynamic reading performance of the UHF RFID system. Figure 3.2a shows the schematic of the temperature control system, Fig. 3.2b shows the real graph and Fig. 3.2c displays the whole schematic diagram of the dynamic test.

A tag is attached to the inner wall of the plastic box. The tag identification rate influenced by the plastic box could be ignored due to the low dielectric constant. A temperature probe hangs inside the plastic box. Simultaneously, some semiconductor heaters are placed at the bottom of the plastic box and connected with a temperature controller outside the plastic box.

First, setting the temperature of the temperature controller, the semiconductor heaters/coolers heat/cool the surrounding temperature inside the plastic box. Second, once the internal temperature reaches the set value, the temperature controller stops supplying power to the semiconductor heaters/coolers.

A temperature controller based on thermoelectric cooler (TEC) was designed here to realize the monitoring of environmental temperature. So the tag identification performance influenced by environmental temperature could be studied.

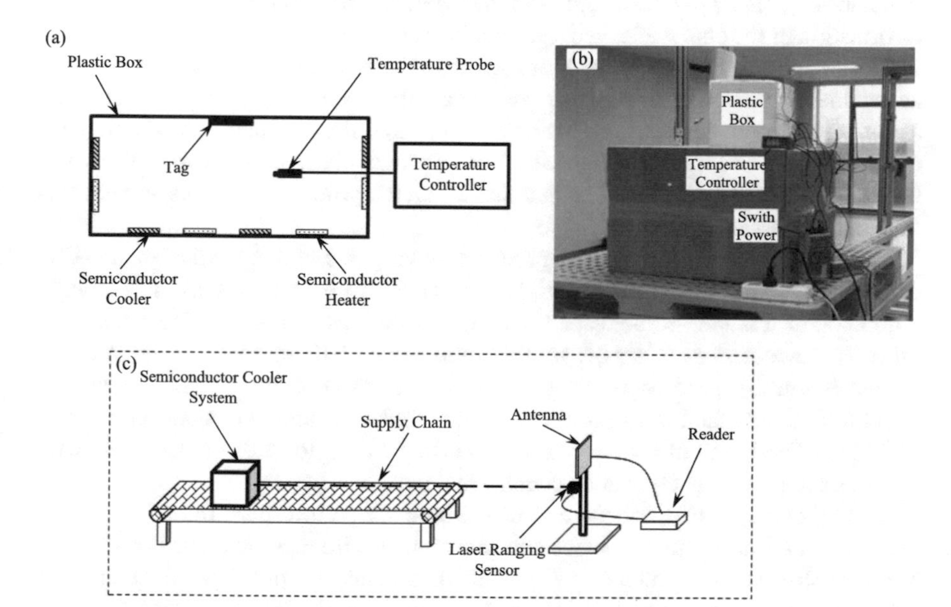

Fig. 3.2 Diagram of temperature control system (color online)

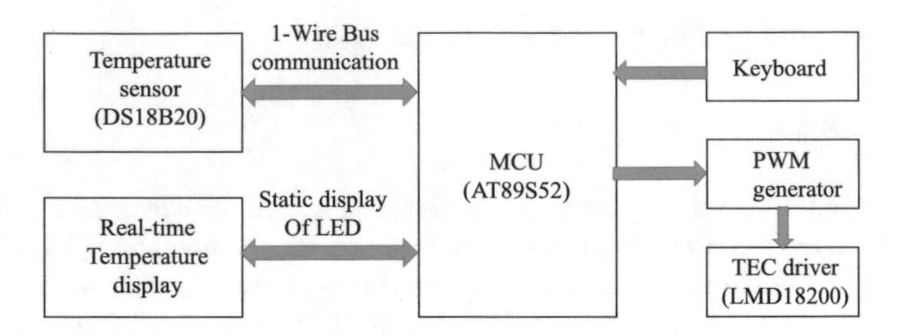

Fig. 3.3 Hardware structure of the temperature-control system

The hardware of the system consists of a thermoelectric cooler (TEC-12702), TEC driver (LMD18200), temperature sensor (DS18B20), digitron, and processor. Figure 3.3 shows a schematic of the system. After starting the system, the microcontroller unit (MCU) measures the environmental temperature, compares the measured value with the set-value, and then outputs the suitable pulse width modulation (PWM) waveform to the TEC driver. Finally, TEC realizes heating and refrigeration. The setting temperature value and the alarm temperature value could be inputted via the keyboard. During the whole course, the MCU repeats this work to keep the environmental temperature at a constant value.

(2) Experimentation platform

In the intelligent supply chain and asset management, RFID tags could hold multiple information of the object attached to, including serial numbers, configuration instructions, and much more. When the shipments arrive in the unloading area, RFID readers at the doors examine their contents and update the inventories of supply chain and asset management accordingly. Inside a warehouse, the products can be automatically identified and tracked wirelessly. Once a product is taken away from the warehouse, the RFID readers installed in doors check the product's contents and updates inventories immediately.

To simulate the environment of products moving in and out, we design an RFID detection system, as shown in Fig. 3.4. The RFID detection system is mainly composed of a reader, some reader antennas, some tags, a laser ranging sensor and other assistive devices. The tags are the commercial UHF tags and the brand of the readers is Impinj speedway revolution R420. The brand of the readers' antennas is Larid A9028 with their frequency band is 902–928 MHz and the center frequency is 915 MHz. The brand of the laser ranging sensor is Wenglor X1TA101MHT88 with its measuring range is 50 m and accuracy is 2 μm.

The following is the procedure of detection in cargo transport. Initially, a tray is installed on a transportation device and some boxes with tags are placed on the tray. Then, an RFID reader and a plurality of RFID antenna are installed on the antenna stand, and the beam of laser ranging sensor points to the boxes. Afterward, the cycle index of the tray is set and the tray transports boxes on the transportation device at a certain speed to simulate the goods in–out warehouse. When the tagged boxes enter

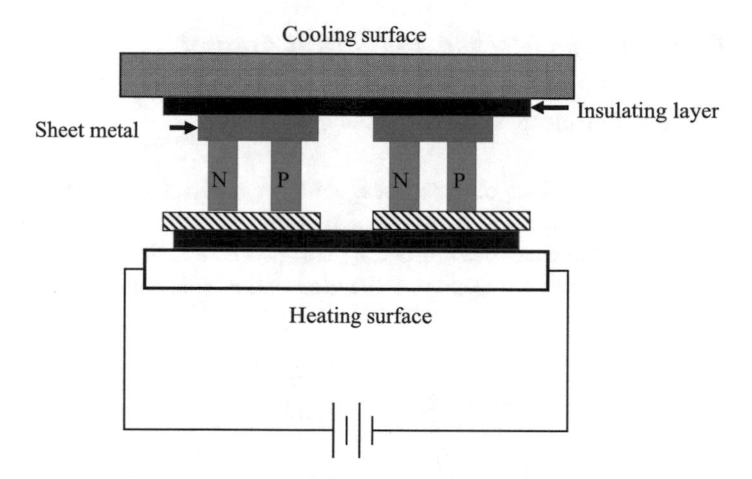

Fig. 3.4 Schematic physical diagram of RFID detection system

the reading zone of the reader antennas, the antennas receive the RF signal of tags and the reader sends a hopping signal to activate the laser ranging sensor. Consequently, we could calculate the reading distance of the RFID antenna to tags. Subsequently, the tray returns to the initial point and repeats the above operation until the cycle index reaches the set value. Eventually, the average value of the measurement is used to be the reading distance of the RFID antenna to RFID tags.

The measurement of reading distance is indirect to the survey. Adjusting the optical lifting platform to ensure the laser beam of the laser ranging sensor aiming at the boxes, we define the intersections of laser ranging sensor's beam and antenna stand plane to reference point. The distance of tags to the reference point is

$$T = S - L \tag{3.18}$$

where L is the deterministic distance of laser ranging sensor to reference point and S is the distance of laser ranging sensor to tags.

The distance of ith RFID antenna to the tag is

$$R_i = (T^2 + H_i^2)^{1/2} \tag{3.19}$$

where Hi is the distance of ith RFID antenna to the reference point. Once the antennas of the reader are fixed on the antenna stand, the distance of ith RFID antenna to the reference point is measured manually and then input to the main program. In the whole measurement, the distance is constant. In practice, the boxes are set to the same and the tags are on the front of the boxes.

3.1.3 Experimental Method and Result Analysis

(1) Influence of the thickness of the plastic box

The transmission coefficient of electromagnetic waves in a material has a great relationship with the dielectric constant of material and the larger the dielectric constant, the lower the transmission coefficient. Moreover, the attenuation speed of an electromagnetic wave depends on the attenuation constant (α) in conductive medium:

$$\begin{aligned} &\alpha = \omega\sqrt{\tfrac{\mu\varepsilon}{2}\left[\sqrt{1+\left(\tfrac{\sigma}{\omega\varepsilon}\right)^2}-1\right]} \\ &\alpha \approx \tfrac{\sigma}{2}\sqrt{\tfrac{\mu}{\varepsilon}} \\ &\sqrt{1+\left(\tfrac{\sigma}{\omega\varepsilon}\right)^2} \approx 1+\tfrac{1}{2}\left(\tfrac{\sigma}{\omega\varepsilon}\right)^2 \end{aligned} \quad (3.20)$$

where ω is the angular frequency of electromagnetic wave, μ is medium permeability, ε is the dielectric constant of the medium. If $\frac{\sigma}{\omega\varepsilon} \ll 1$:

$$\sqrt{1+\left(\frac{\sigma}{\omega\varepsilon}\right)^2} \approx 1+\frac{1}{2}\left(\frac{\sigma}{\omega\varepsilon}\right)^2 \quad (3.21)$$

Hence, Eq. (3.20) could be rewritten as

$$\alpha \approx \frac{\sigma}{2}\sqrt{\frac{\mu}{\varepsilon}} \quad (3.22)$$

Therefore, we use the polytetrafluoroethylene (PTEE) to make a plastic box, which has a small dielectric constant. Supposing the relative dielectric constant $\varepsilon_r = 2.1$, the relative magnetic permeability $\mu_r = 1$, and bulk conductivity $\sigma_V = 2.5\times10^{-17}$ S/cm, we could obtain the effects of medium thickness on attenuation constant, as shown in Fig. 3.5.

From Fig. 3.5, the attenuation constant increase gradually with the increase of the medium's thickness, but the index is 10^{-15}, so the attenuation of the electromagnetic wave in the medium can be ignored.

In order to verify the influence of the plastic box's thickness on the reading performance of the RFID tag, we test the plastic boxes at room temperature, whose thickness is 0.3, 0.5, 1, 2, and 3 mm respectively, as shown in Fig. 3.6.

The thickness of the plastic box has little effect on the performance of tags in the range of 0.2–3 mm and the maximum range is 0.02 m, which could be considered within the limit of measurement error in practical applications. Simultaneously, the error is equal for the reading distance at different temperatures, so the following

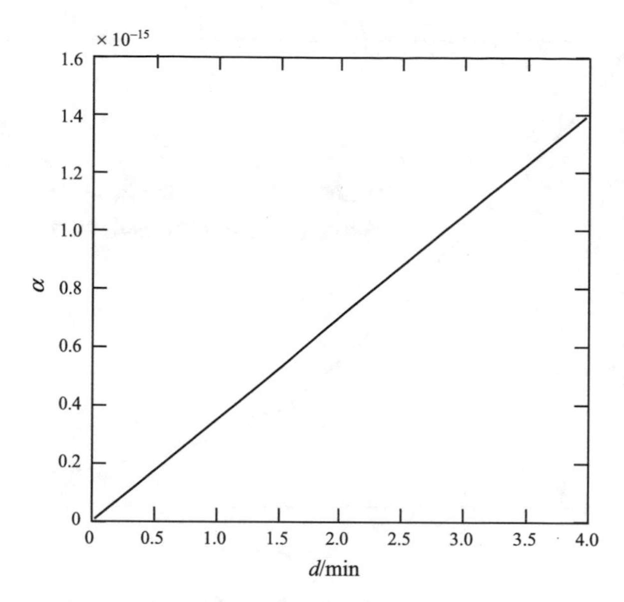

Fig. 3.5 Effects of medium thickness on attenuation constant

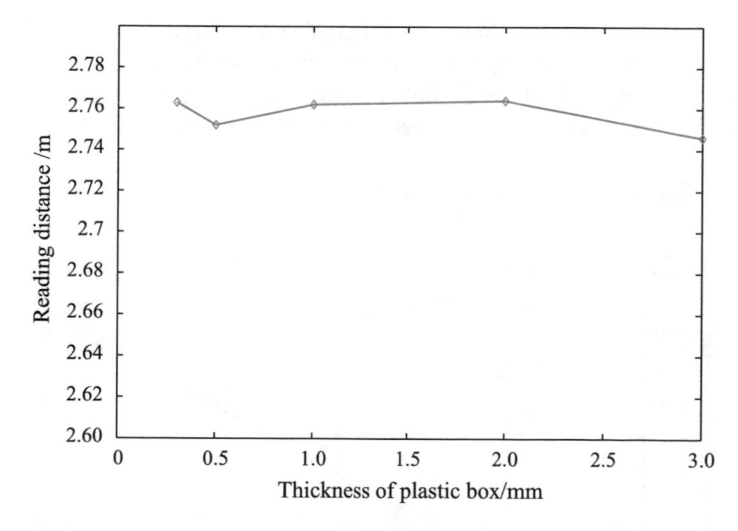

Fig. 3.6 Influence of the plastic box's thickness on the reading distance of RFID tag

measurement results are the ones after subtracting the noise. For convenience, we choose the plastic box in 3 mm to test at different temperatures.

The reading distances of the tag are measured in 13 sets of temperature from 20 °C to 80 °C. Then, a scatter diagram could be obtained by analyzing the measured data, as shown in Fig. 3.7. The horizontal coordinate stands for test temperature and the

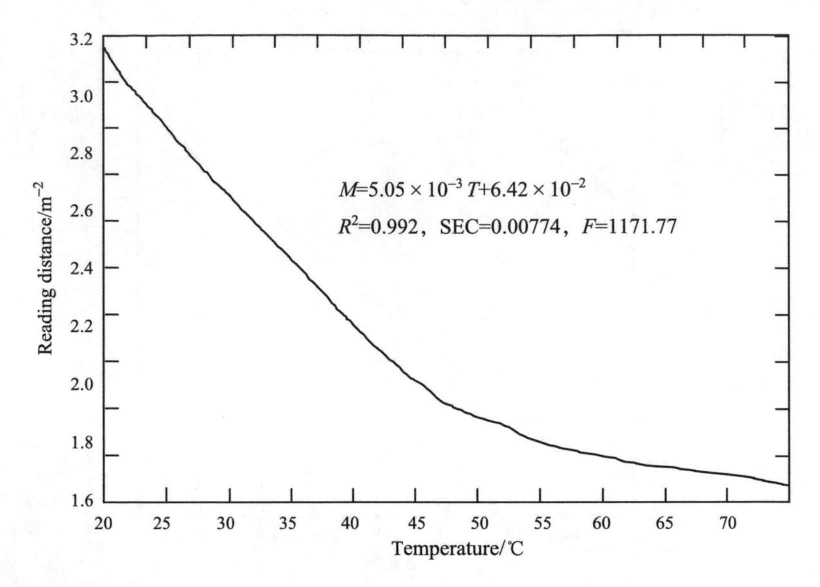

Fig. 3.7 Scatter diagram and fitting line of temperature and reading distance

vertical coordinate represents the tags' reading distance. Equation (3.15) shows that the test temperature is directly proportional to reciprocal reading distance, so a scatter plot is shown in Fig. 3.8, in which the horizontal coordinate represents test temperature and the vertical coordinate represents the tags' reciprocal reading distance.

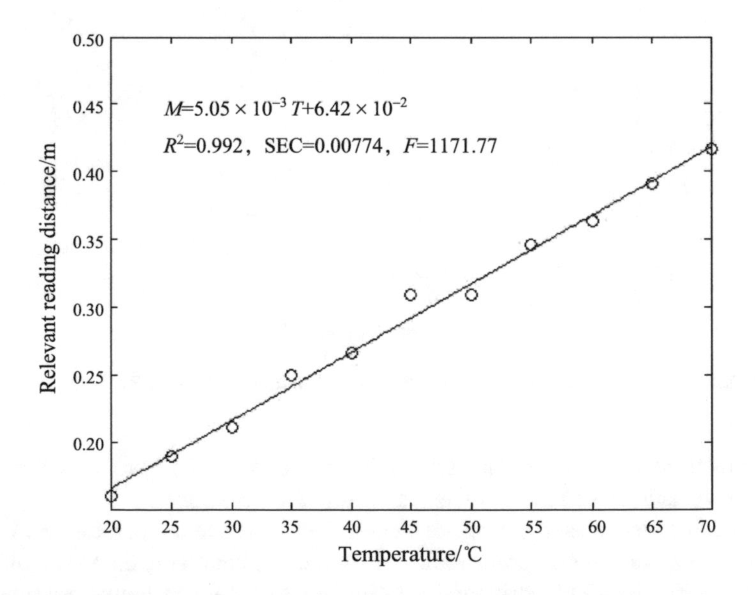

Fig. 3.8 Scatter diagram and fitting line of temperature and relevant reading distance

The correlation coefficient R is an index to measure the linear correlation between two random variables. R^2 is the determination coefficient. It represents how good the correlation between the two random variables is. The larger these two coefficients are, the better the correlation between the two variables is. From Fig. 3.8, we can see that R^2 is more than 0.95, which means a good linear correlation between alcohol concentration and corrected correlation coefficient. The standard error of calibration SEC is a key to evaluate the quality of the model equation. The smaller the value of SEC is, the better the equation is. From Fig. 3.8 we can see that the value of SEC is very small, so the calibration error of the equation is also small. F is the significance of the model and the linear relationship of the regression equation is significant while F is large. In summary, the obtained model equation Eq. (3.12) has a good correlation and a significant linear relationship.

(2) Predication of the reading distance of different temperatures

In order to verify the accuracy of the measurement method, ten different temperatures are prepared. The designed system is used to measure different R_i and calculate different values of the RFID tag's reading distance at each temperature. Then we could predict the reading distance of the RFID tag at different temperatures by the model equation Eq. (3.16). The experimental data of the test temperatures are shown in Table 3.1.

The calculated data and errors of test temperatures are shown in Table 3.1, and the evaluation parameters of the model equation of the prediction experiment are shown in Table 3.2. The scattering diagram of the prediction reading distance is shown in Fig. 3.9, where X-axis represents the reference reading distance and Y-axis represents the prediction reading distance. In Table 3.2, the correlation coefficient r_p of the model equation is close to 1, which means the relationship between prediction

Table 3.1 Experimental data of the test temperatures

Samples	Temperature T/°C	Reading distance R/m	Prediction of relevant reading distance M/(1/m^2)	Prediction of Reading distance R/m	Error /%
1	22	2.42	0.175	2.39	1.24
2	26	2.28	0.196	2.26	0.88
3	32	2.12	0.226	2.10	0.94
4	37	2.01	0.251	1.99	0.99
5	43	1.86	0.281	1.89	1.61
6	47	1.83	0.302	1.82	0.55
7	52	1.73	0.327	1.75	1.16
8	59	1.68	0.362	1.66	1.19
9	63	1.60	0.382	1.62	1.25
10	67	1.56	0.403	1.58	1.28

Table 3.2 Prediction evaluation parameters of the model equation

r_p	0.9538
SEP	0.0229

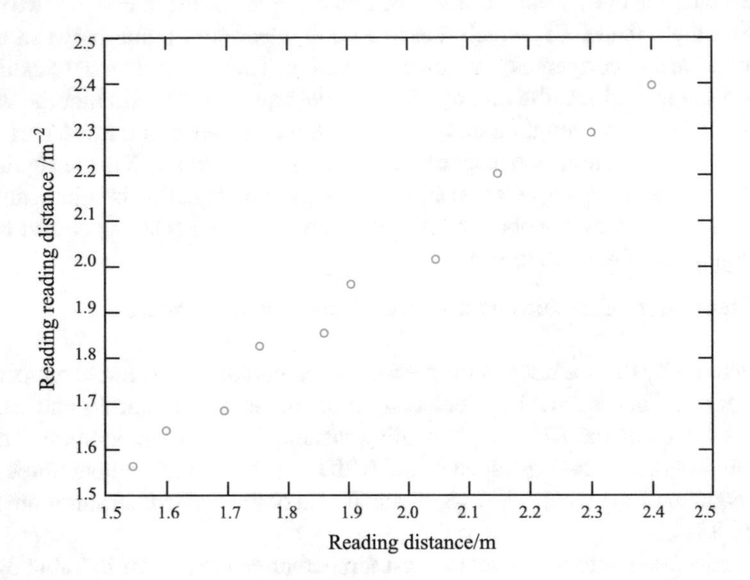

Fig. 3.9 Scatter diagram of prediction reading distance

reading distance and the corrected correlation coefficient is significant. And the standard error of prediction SEP has a very small value. From Table 3.2, we can see that the error range of the measurement is within ±3% and the main reason is that different temperature have a slight influence on reflecting and absorbing an electromagnetic wave of the plastic box.

In practice, wireless communication channels are subject to channel impairments such as fading in addition to additive noise. Similarly, the reading distance of tags is greatly influenced by the noise in the channel, particularly fading, which could cause frequency offset [16]. In the case of frequency offset, the influence of fading on reading distance could be given via replacing f' with $f'' = (f' + \Delta f')$ in Eq. (3.13), in which $\Delta f'$ is the offset of working frequency. This work is in the presence of assuming no fading to consider the effects of temperature on the reading performance of tags at low speed in the indoor environment.

3.2 Electromagnetism Analysis of Physical Anti-Collision—Research on the Effect of Humidity and Salt Fog on Tag Performance

3.2.1 Effect of Humidity on the Moving Identifying Performance of UHF RFID System

Passive UHF RFID systems have a longer read range than other RFID systems. Therefore, They are widely used in transportation fields including road transportation and ship transportation. Danaela [17] used RFID loop tags on an onboard ship. For cost reasons, Danaela thought that product identification should use active readers and passive tags. Considering ship locks for ships transporting in rivers, Hu et al. [18] proposed a ship access lock management system by using an RFID system, which was used to complete the tracking work during the operation of the ship. This method completed the ship non-stop by the lock-in procedure in the river, which made service more convenient and had broad prospects. Through the RFID upgrade and modification of existing containers, in and out efficiency of the port can be greatly improved [19]. The proportion of today's social transportation industry is constantly increasing. Through the RFID upgrade of traditional storage systems and container yards, transportation efficiency can be greatly improved. With the help of internet of things technology, the seaport will be greatly promoted as a benchmark for high-volume transportation. Meanwhile, the problems of the effect of humidity on the UHF RFID system's performance keep springing up.

The RFID system can be applied in extremely harsh environments such as liquid and metal environments. Through a lot of experiments, Wu [20] found that higher conductivity liquid may have a greater impact on RFID tag's reading performance, but he did not theoretically analyze the reasons. How to ensure the integrity of equipment is an important issue in modern industry. Liquid leak detection can effectively ensure the safety of instruments and equipment. Guo et al. [21] detected liquid leakage by using COTS RFID devices. Periyasamy et al. [22] analyzed the 13.56 MHz passive RFID tag's performance in liquid and metal environments. The purpose of the study was to investigate whether particularly close access to different types of metals and liquids would have an effect on the RFID system's reading performance. The reading rate and the reading distance between tag and reader were used as performance criteria to test the changes of the system under different influencing factors.

The RFID system will have great development potential in the medical field in the future. RFID tags in medicine are often used in a liquid-related environment or harsh environment. Ergin et al. [23] used RFID tags for pharmaceutical products. Distance between RFID tag and interrogator, dosage form, rotation angle, and ion concentration in solution were used as influencing factors in the study. The results showed that any dosage form could affect the tag compared to an empty container. The specific impact was the decrease in read performance. Alfonso et al. [24] attached RFID tags to the blood bag. RFID tag's viability was tested in harsh environments to simulate the environment of a blood bank. The research demonstrated that the reading ability of RFID tags could be affected under extreme conditions.

Due to the advantages of the UHF RFID system, it can be used in special environments, including high temperature or high humidity environments like deserts and lakes. Yu et al. [25] designed a testing platform for the evaluation of UHF RFID tags' performance at different temperatures. Whether the harsh environments will have an influence on the tag's performance has become an important issue in tag improvement research. In past studies, it has been analyzed from the perspective of temperature influence, but few studies have been done from a humidity perspective. In this chapter, the semi-physical simulation of the humidity environment simulates the gradual process from low humidity to high humidity. Furthermore, the humidity influence theory is proposed, which is of great significance for the study of UHF RFID tag performance.

(1) Fundamental principle

Electromagnetic waves are emitted from the antenna to the surrounding space and encounter different targets. Part of the electromagnetic energy reaching these targets is absorbed and the other part is scattered in various directions with different intensities. A portion of the reflected energy is eventually returned to the transmitting antenna.

For reflective scattering RFID systems, the tag uses electromagnetic reflected waves to complete the energy transfer from the tag to the reader. Its energy transmission process can be divided into forward link and reverse link transmission.

The power density of the electromagnetic waves incident on the RFID tag antenna in free space is

$$S = \frac{P_t \times G_t}{4\pi R^2} = \frac{\mathrm{P}_{\mathrm{EIR}}}{4\pi R^2} \tag{3.23}$$

where

P_t is the transmit power of the reader,
G_t is the gain of the reader transmit antenna,
R is the distance between the electronic tag and the reader, and
P_{EIR} is the effective radiated power of the transmit antenna.

When the maximum radiation direction of the tag antenna is consistent with the reader antenna and the polarization of the two are matched, the tag antenna can absorb the maximum power from the electromagnetic wave proportional to the power density of the incident wave:

$$P_{\mathrm{tag}} = A_{e\_tag} S = \frac{\lambda^2 G_{\mathrm{tag}} S}{4\pi} = P_{\mathrm{EIR}} G_{\mathrm{tag}} \left(\frac{\lambda}{4\pi R}\right)^2 = P_t G_t G_{\mathrm{tag}} \left(\frac{\lambda}{4\pi R}\right)^2$$

$$y[n] = \alpha A(\theta) s[n] + w[n] (n = 1, 2, \ldots N)$$

$$CRB(\theta) = \frac{1}{2SNR\left(Ma^H(\theta)R_S^{\mathrm{T}}a(\theta) + a^H(\theta)R_S^{\mathrm{T}}a(\theta)\|a(\theta)\|^2 - \frac{M\left|a^H(\theta)R_S^{\mathrm{T}}a(\theta)\right|^2}{a^H(\theta)R_S^{\mathrm{T}}a(\theta)}\right)}$$

$$CRB(\theta) = \frac{1}{8NM\frac{|\alpha|^2}{\sigma^2}\left(\sum_{k=-(M-1)/2}^{(M-1)/2} k^2\right)(\pi\cos\theta)^2(d_b^2 + d_a^2)}$$

$$R_S = \begin{bmatrix} 1 & \beta_{12} & \cdots & \beta_{1M} \\ \beta_{21} & 1 & \cdots & \beta_{2M} \\ \vdots & \vdots & & \vdots \\ \beta_{M1} & \beta_{M2} & \cdots & 1 \end{bmatrix} \quad (3.24)$$

where

G_{tag} is the gain of the tag antenna,
$A_{\text{e\_tag}}$ is the effective area of the tag antenna.

In UHF RFID systems, the electromagnetic waves emitted from the reader provided the energy required for the tag circuit to work properly, so the power consumption of the tag directly affects the reading and writing distance. Conversely, for the same tag, to increase the read and write distance, we need to increase the reader's transmit power.

In electromagnetic field theory, the energy emitted by a tag is proportional to the radar cross section of the tag's antenna σ, and σ is an important parameter to measure the ability of a target to reflect electromagnetic waves. The radar cross section is affected by many factors, including electromagnetic wave wavelength, target size, shape, material. Tag reverse reader reflects electromagnetic energy, P_{Back} is given by

$$P_{\text{Back}} = A_{\text{e\_reader}} S_{\text{Back}} \quad (3.25)$$

The electromagnetic wave's power density backscattered by the tag to the reader S_{Back} is

$$S_{\text{Back}} = \frac{P_T G_T \sigma}{(4\pi R^2)^2} \quad (3.26)$$

The effective area of the reader receiving antenna is

$$A_{\text{e\_reader}} = \left(\frac{\lambda^2}{4\pi}\right) G_r \quad (3.27)$$

where

G_r is the gain of the reader to receive the antenna.

When the reader transmits and accepts the same antenna, there is $G_r = G_t$. Therefore, the power received by the reader's antenna position is

$$P_{\text{Back}} = P_t G_t G_r \lambda^2 \frac{\sigma}{(4\pi)^3 R^4} \tag{3.28}$$

Generally, the forward link limits the achievable reading range in an RFID system because the sensitivity of the reader is significantly higher than the sensitivity of the tag chip. Tag's maximum read distance is given by

$$d = \frac{\lambda}{4\pi}\sqrt{\frac{P_t G_r G_t \tau}{P_{th}}} \tag{3.29}$$

As the ambient humidity increases, so does the power required to activate the tag which means it will change the tag chip sensitivity threshold P_{th} [26]. Moreover, the tag chip sensitivity threshold P_{th} shift is a result of humidity variation. Previous research shows chip sensitivity threshold P_{th} has a linear relationship with humidity. Let us assume that, with the increase of humidity, the tag chip sensitivity threshold P_{th} increased gradually.

$$\begin{aligned} d &= \frac{\lambda}{4\pi}\sqrt{\frac{PtGtGr\tau}{Pth'}} \\ &= \frac{\lambda}{4\pi}\sqrt{\frac{PtGtGr\tau}{10\,\text{lg}(ax+b)}} \end{aligned} \tag{3.30}$$

When the parameters of the tag are determined and the signal strength value reaches the activation threshold, the tag can be read. Equation (3.30) can be rewritten as

$$d = \eta \frac{1}{\sqrt{10\,\text{lg}(ax+b)}} \tag{3.31}$$

$$\eta = \frac{\lambda}{4\pi}\sqrt{PtGtGr\tau} \tag{3.32}$$

In this case, η could be considered as a constant in the simulation.

Let P_t. $G_t = 33\text{dBm} = 1995\text{mW}$, $G_r = 2\text{dBi}$, $a = 0.00186$, $b = 0.9413$, $\tau = 1$.

Previous research shows that if the distance between the surface and the tag of the box is $\lambda/4$ [27], the tag's reading rate has the best effect. If the removal from the tag to the test chamber's surface is $\lambda/4$, previous research shows that the transmission coefficient $\tau \approx 1$ [28]. Finally, we can obtain the connection between the tag's reading distance and the humidity under static conditions, as shown in Fig. 3.10.

(2) Design of the Experimentation System

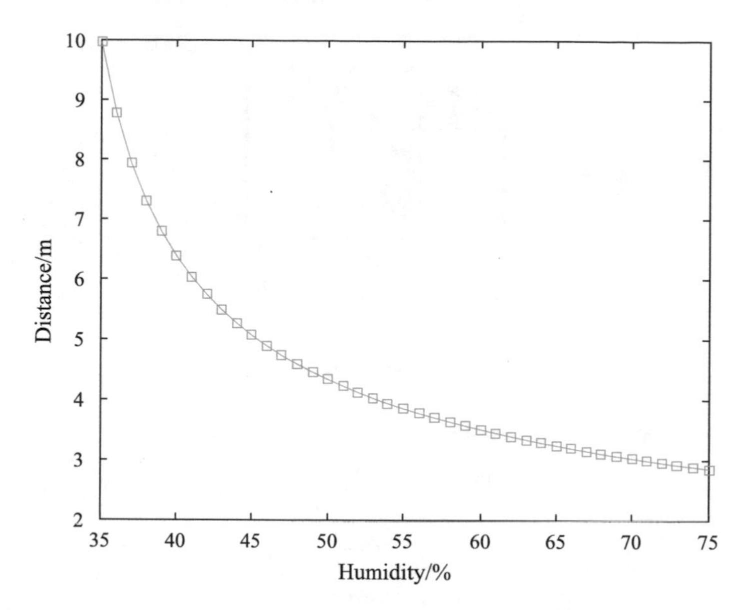

Fig. 3.10 Connection between the tag's reading distance and the humidity under static conditions

The constant humidity moving experiment detection system is shown in Fig. 3.11. The humidity test chamber is suitable for measuring and controlling environmental humidity. The system simulates the effect of humidity on the UHF RFID tag's reading properly. Figure 3.11a shows the schematic of the humidity test chamber, Fig. 3.11b shows the schematic of the moving experiment system, Fig. 3.11c shows a physical diagram of the humidity experiment box, Fig. 3.11d shows the hardware structure of a humidity-control system, Fig. 3.11e shows the structure of the humidity experiment box.

The test chamber's inner wall affixed the RFID tag. As the dielectric constant is very low, the tag recognition rate affected by plastic boxes can be ignored. The humidity probe is suspended in the test chamber. Some heaters and humidifiers, connected to a humidity controller outside the test chamber, are placed on the bottom of the test chamber.

We set the experimental humidity in the humidity controller. The heater and the humidifier control the surrounding humidity inside the test chamber. The humidity controller will stop the heater and humidifier's power supply, if the humidity in the test chamber reaches the set value.

The hardware of the system consists of these parts. We choose HTC-201 humidity controller as the humidity controller, HT211 humidity probe as the humidity probe, JRD type aluminum alloy heating sheet as the dehumidifier, ultrasonic humidifier as the humidifier.

RFID tags can be affixed to the surface of the cargo when it is received at the port, where ambient humidity changes greatly. Due to the reliability of RFID tags, a lot of information can be stored in the tags such as product shipping information

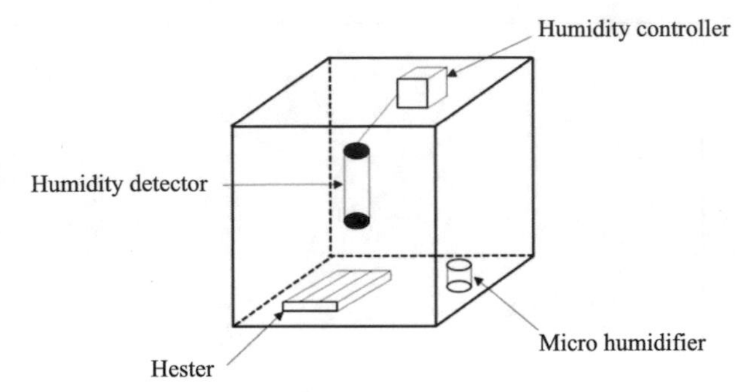

(a) Schematic of the humidity test chamber

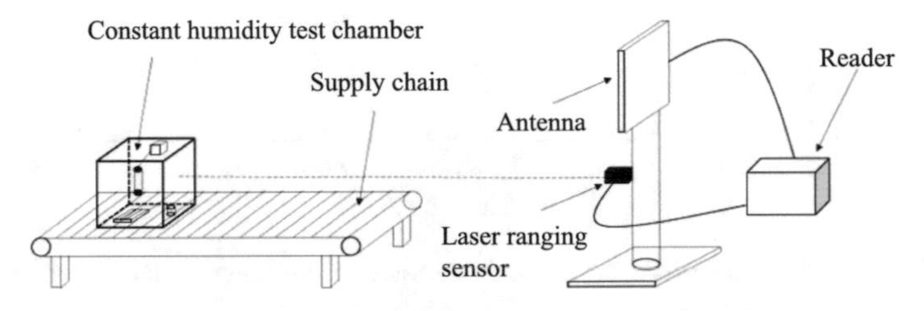

(b) Schematic of the moving experiment system

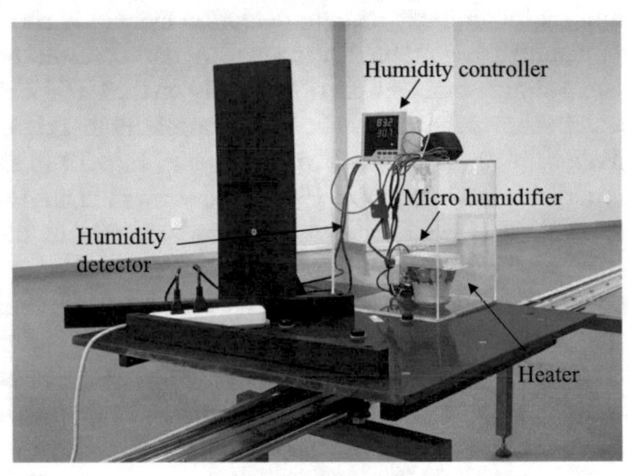

(c) Real humidity test chamber

Fig. 3.11 Humidity control system

and product attributes. When the goods with RFID tags reach the unloading area, the goods can be automatically identified and goods' information will be transmitted to the computer for storage.

We designed an experimental platform to simulate the process of goods with RFID tags entering and leaving the warehouse under different humidity, which is shown in Fig. 3.12. The experimental platform is constructed by the following parts:1. reader, 2. reader antenna, 3. laser ranging sensor, and 4. other auxiliary equipment.

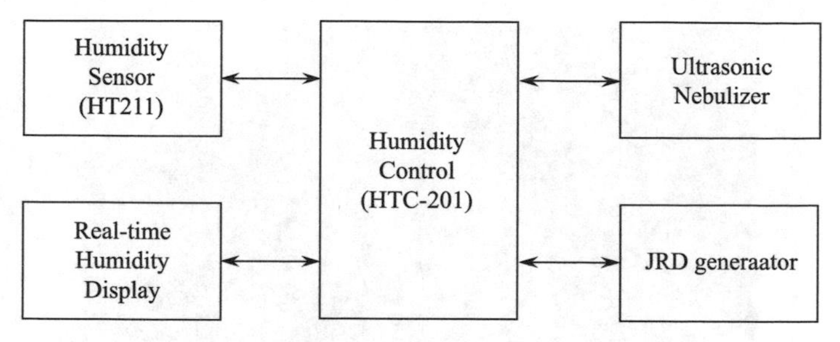

(d) Hardware structure of a humidity-control system

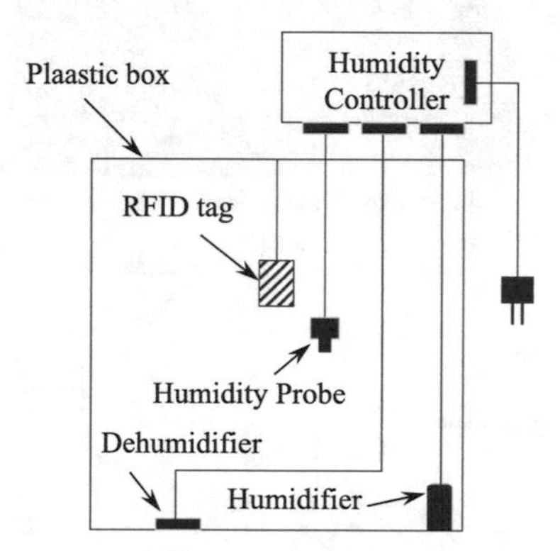

(e) Structure of the humidity experiment box

Fig. 3.11 (continued)

The dynamic measurement system's RFID antenna uses Larid A9028 far-field antenna. Antenna's maximum reading range is about 12 m. We use the Speedway Revolution R420 UHF reader as the RFID reader. Distance measuring sensor is X1TA101MHT88 laser distance measuring sensor from Germany Wenglor.

The following are the detection steps in cargo transportation under different ambient humidity. First, the pallet is mounted on a transport device, and a box with an RFID tag is put on the pallet. Then, an RFID reader and a plurality of RFID antennas are mounted on the antenna frame. Laser rangefinder aimed at the experiment box. Finally, we set the conveyor speed to simulate the cargo unloading process.

If the tagged product enters the RFID tag reader's reading range, the antenna emits electromagnetic waves to activate the tag, then the tag reflects the electromagnetic waves to transmit the information in the tag, and the reader activates the laser ranging radar. Therefore, the distance from the antenna to the tag can be calculated as the distance of the product from the antenna.

(3) Experimental Method and Result analysis

We chose three types of box thicknesses for testing. The data is shown in Table 3.3. The data shows that the impact of different thickness plastic test chamber on the RFID tag's reading range is negligible. Considering the test chamber's cabinet

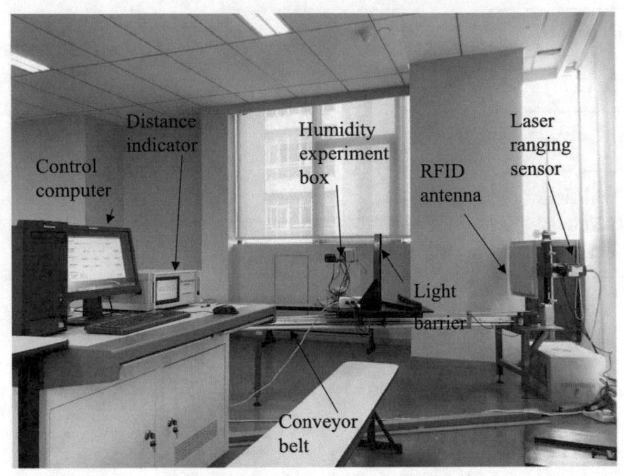

(a) Real experimentation platform

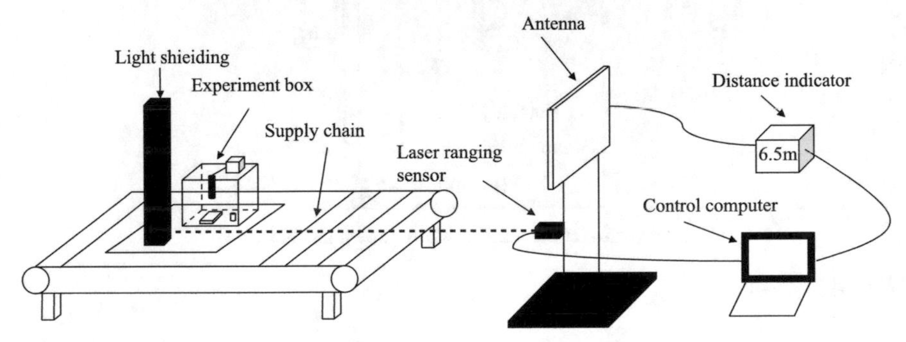

(b) Component diagram of RFID test platform

Fig. 3.12 Assembly drawing of RFID detection system

Table 3.3 Test results of test chamber's thickness

Thickness of plastic box/mm	Reading distance/m
0.5	6.030
1.0	6.021
1.5	6.002
2.0	5.986
2.5	5.971
3.0	5.962

strength, we chose the test chamber with a thickness of 3 mm. The thickness of the test chamber can hardly affect the reading range of the RFID tag.

The height of the RFID tag from the ground in the range of 0.7–1.3 m has a great impact on the property of the RFID tag. Since the conveyor height and antenna height are fixed, the antenna has a reading range. To test the optimal ground clearance for

RFID tags, we chose 7 heights. The data is shown in Table 3.4. According to the reading distance and reading rate, the RFID tag height of 1 m was selected in the experiment.

Tag's reading distance is measured in 41 different humidity ranges from 35 to 75%. Data of the experimental reading range and theoretical curve for each humidity is shown in Fig. 3.13. In Fig. 3.13, the vertical coordinate represents the reading range of the tag and the horizontal coordinate represents the test humidity. The stars represent

Table 3.4 Experimental data of the RFID tag height from ground

Height/m	Antenna power/dBm	Average reading distance/m	Test times	Successful readings	Read rate/%
0.7	28	6.153	10	9	90
0.8	28	7.4051	10	10	100
0.9	28	4.78	10	8	80
1.0	28	12.977	10	10	100
1.1	28	3.485	10	10	100
1.2	28	2.465	10	9	90
1.3	28	11.8924	10	10	100

Table 3.5 Evaluation index of the model

R-square	0.9406
Average error	4.13%

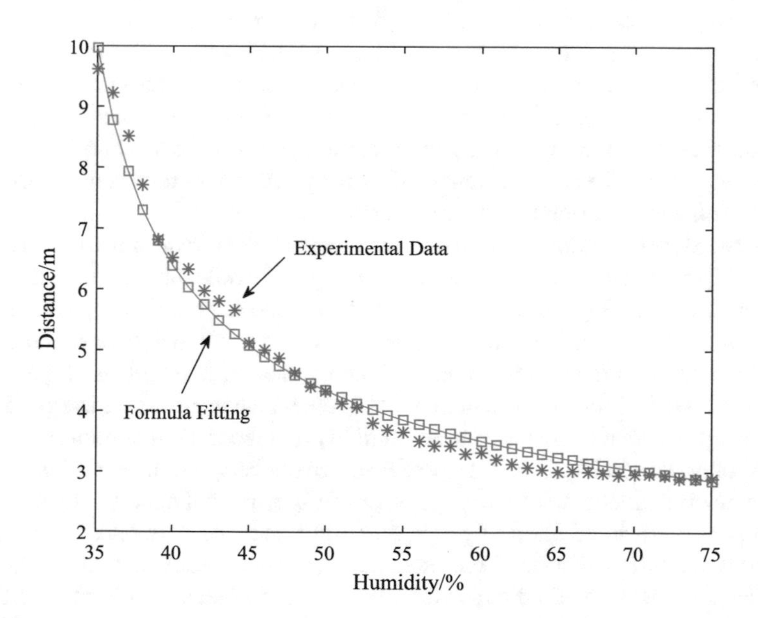

Fig. 3.13 Experimental data of the test humidity and theoretical curve

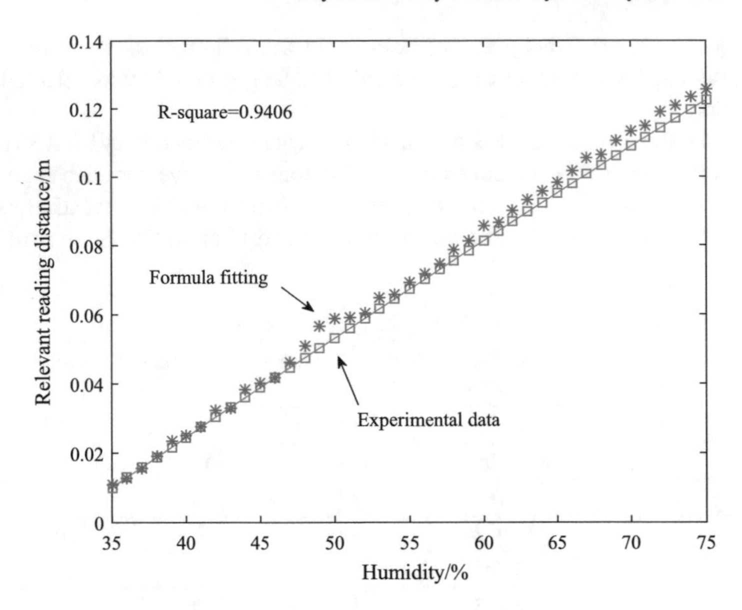

Fig. 3.14 Scatter plot and fitted line for humidity and related reading range

the experimental data. The squares represent formula fitting. The experimental data is consistent with the theoretical trend.

The system measures different reading distances of RFID tags at each humidity value. By evaluating and verifying the model, the obtained humiditysimulates the reading range of the tag under different humidity. From Eq. (3.31), it can be seen that the experimental humidity has a reciprocal relationship with the tag's reading distance. By using the relevant reading distance, the accuracy of the predictive model can be verified and formula calculation can be simplified. The scatter plot of the experiment is shown in Fig. 3.14. The abscissa represents the test humidity, and the ordinate represents the tag's relevant reading range. Asterisks represent experimental data and squares represent the theoretical curve.

R is called the correlation coefficient, which is an indicator that measures the linear relationship between two random variables. *R*-square is called the coefficient of determination. Determining coefficients indicate how good the relationship between two random variables is. The larger the coefficient of determination is, the better the correlation between two random variables is. The evaluation index of the experimental model is shown in Table 6.5. *R*-squared is 0.9406, which means there is a good linear relationship between the experimental humidity and theoretical humidity.

It is found that in Fig. 3.13 there are some errors between the experimental data and the theoretical curve. The average error of experimental data is 4.13%. In practical applications, in addition to generating additional noise, wireless communication channels will also suffer from channel impairments such as fading. Humidity increases the power required to activate the tag which means it will change the tag chip sensitivity threshold P_{th}. At the same time, the reading range of the tag will also

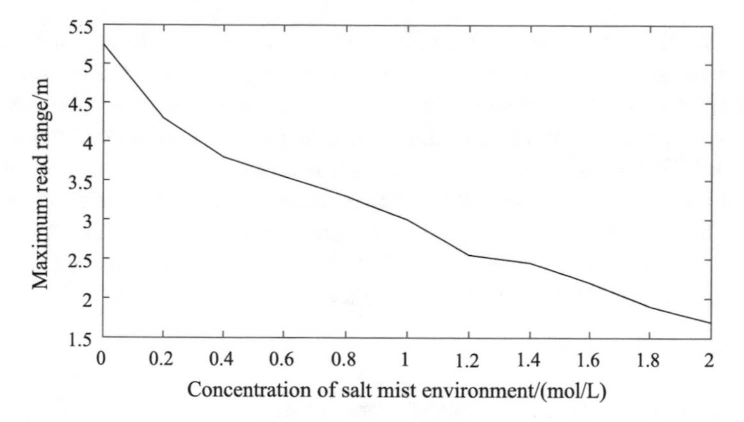

Fig. 3.15 The maximum reading distance in simulation salt mist environment

be significantly affected by the noise (especially fading) in the communication. In this case, a new reading range model could be given by using $P_{th'} = (P_{th} + \Delta P_{th})$ in Eq. (3.31). Other errors may be caused by changes in ambient temperature or interference from other electronic instruments in the environment. Future work will carry out corresponding in-depth research with a higher recognition rate and stability, as well as incorporating other environmental factors.

3.2.2 *Effect of Salt Mist on the Moving Identifying Performance of UHF RFID System*

Due to the globalization of trade, the proportion of marine transportation in total trade is increasing, and RFID-based tags do not require contact measurement. This shows the importance of carrying out NaCl environmental testing. Therefore, some scholars have recently studied the impact of NaCl environment on the performance of RFID systems. Barge et al. pointed out that some compounds, such as NaCl, have a significant impact on RFID reading performance and tested the minimum power required to activate several different types of labels at various NaCl solution concentrations [29]. Potyrailo et al. integrated RFID sensors into single-use biopharmaceutical manufacturing components, and tested these sensors for long-term stability of temperature and solution conductivity, examined the response of RFID conductivity sensors to several common buffers including sodium chloride (NaCl). It shows that the sensor number enhances with the increase of NaCl concentration [30]. Ennasar et al. founded that the conductivity and dielectric constant of water affect the performance of RFID tags when NaCl or sucrose is added. For all frequencies in the UHF RFID band, the response frequency of the tag antenna increases with the concentration of sucrose and NaCl in the water [31].

However, little research has been done on the dynamic identification of RFID in the salt mist environment. In this chapter, a semi-physical experiment chamber for simulating salt mist environment is designed, which simulates a gradual process from low to high concentration of salt mist environment in 80% humidity environment. Then, a RFID dynamic identification platform based on salt mist environment is designed and built. The system gets the distance between the reader antenna and the tags in salt mist environment with different concentrations. The variation of conductivity in different concentrations in salt mist environment impacted the RFID recognition distance is studied. The theoretical derivation is given.

(1) Fundamental principle

Electromagnetic waves are transmitted from the antenna to the surrounding space to encounter UHF RFID electronic tags. A part of the electromagnetic energy reaching the electronic tag is absorbed, and the other part is reflected in different directions with different strengths. Some of them eventually return to the transmitting antenna, and the system reads the tag information. The change of NaCl concentration in salt mist environment will affect the conduction of electromagnetic waves. The change of humidity will affect the threshold power of electronic tags. We will analyze them one by one.

In our study, the NaCl solution in the salt mist environment is uniformly isotropic. In order to form the salt mist environment with different concentrations, we chose different concentrations of NaCl solutions to be used in the experimental box. The relationship between conductivity and concentration of NaCl solution under 18°C is shown in Table 3.6 [32].

The communication between the reader and the tag is established by backscattering electromagnetic waves in the far field. In our study, the salt mist environment in test chamber is homogeneous isotropic, with dielectric constant is ε, permeability is μ, and conductivity is σ. According to the general wave equation of electromagnetic wave propagation [33], when electromagnetic waves enter the salt mist environment from the air, the electric field in the salt mist environment is:

Table 3.6 Relationship between conductivity and concentration of NaCl solution

Concentration/(mol/L)	Conductivity/(S/m)
0.001	0.011
0.005	0.052
0.010	0.120
0.050	0.479
0.100	0.920
0.500	4.045
1.000	7.430
2.000	12.96

$$E'' = E_0''e^{\mathrm{i}\left(\vec{k}''\cdot\vec{x}-\omega t\right)} = E_0''e^{-\alpha z}e^{\mathrm{i}(\beta z-\omega t)} \tag{3.33}$$

where E_0'' is the electric field amplitude in salt mist environment,ω is angular frequency, α is attenuation coefficient, β is phase coefficient, given by

$$\alpha = \beta = \sqrt{\frac{\omega\mu\sigma}{2}} \tag{3.34}$$

In the RFID testing system, free charge movement in the salt mist environment forms the eddy current, which generates Joule heat. The current density j in the salt mist environment can be given as:

$$j = \sigma E'' \tag{3.35}$$

The average power consumption per unit volume of the salt mist environmentis:

$$\begin{aligned} p &= \frac{1}{2}R_E\left(j\cdot E''^{*}\right) \\ &= \frac{1}{2}R_E\left(\sigma E''\cdot E''^{*}\right) \\ &= \frac{1}{2}R_E\left[\sigma E_0''e^{-\alpha z}e^{\mathrm{i}(\beta z-\omega t)}\cdot E_0''e^{-\alpha z}e^{-\mathrm{i}(\beta z-\omega t)}\right] \\ &= \frac{1}{2}\sigma E_0''^{2}\mathrm{e}^{-2\alpha z} \end{aligned} \tag{3.36}$$

The height of the salt mist environment is set as a, the power per unit surface area is:

$$P = \int_0^a p\mathrm{d}s = \frac{1}{2}\sigma E_0''^{2}\int_0^a e^{-2\alpha z}\mathrm{d}z \tag{3.37}$$

where

$$\int_0^a e^{-2\alpha z}\mathrm{d}z = -\frac{1}{2\alpha}\left(e^{-2\alpha a}-1\right) \approx \frac{1}{2\alpha} \tag{3.38}$$

Therefore, the power of per unit surface area can be deduced as:

$$P = \frac{\sigma E_0''^{2}}{4\alpha} \tag{3.39}$$

By Fresnel formula and the boundary condition equations, transmission coefficient T at the interface of two media is:

$$T = \frac{E_0''}{E_0} = \frac{2\eta_2}{\eta_2 + \eta_1} \tag{3.40}$$

where η_1 and η_2 are the intrinsic impedance, given by $\eta_2 = \sqrt{\mu_2/\xi_2}$, $\eta_1 = \sqrt{\mu_1/\xi_1}$. ξ_1 is the air dielectric constant.ξ_2 is the dielectric constant of salt mist environment.μ_1 is air permeability.μ_2 is the permeability ofsalt mist environment.

The average incident power density $\overline{S}$ is:

$$\overline{S} = \frac{1}{2}\mathrm{Re}\left[EH^*\right] = \frac{E_0^2}{2\eta_1} \tag{3.41}$$

The distance from the surface of the salt mist environment to tag is R. The average power density of the electromagnetic waves on the RFID tag antenna at the surface is:

$$\overline{S} = \frac{P_{\mathrm{re-radiated}}}{4\pi R^2} = \frac{KP_aG_r}{4\pi R^2} \tag{3.42}$$

Combining Eq. (6.41) and Eq. (6.42), E_0^2 can be deduced as:

$$E_0''^2 = \frac{2\eta_1\eta_2^2 KP_aG_r}{(\eta_1 + \eta_2)^2 \pi R^2} \tag{3.43}$$

The salt mist environment test chamber is square with a side length of a. Combine Eqs. (3.34) and (3.39), total power consumption on the surface of salt mist environment W_e can be deduced as:

$$W_e = \frac{\sigma E_0''^2}{4\alpha} * 2a^2 = \frac{\sigma E_0''^2}{4\sqrt{\frac{\omega\mu\sigma}{2}}} * 2a^2 \tag{3.44}$$

When the tag is in the salt mist environment, the salt mist will cause the reduce of tag antenna gain $G_{r'}$. $G_{r'}$ can be deduced as:

$$G_r' = \frac{G_r(P_{\mathrm{re-radiated}} - W_e)}{P_{\mathrm{re-radiated}}} \tag{3.45}$$

Under normal circumstances, the sensitivity of the reader is significantly higher than that of the tag chip, resulting in that the forward link limits the achievable read range in RFID systems [34]. The maximum reading distance is given by:

$$d = \frac{\lambda}{4\pi}\sqrt{\frac{P_t G_r G_t \tau}{P_{th}}} \tag{3.46}$$

where λ is the carrier-frequency wavelength. P_t is the transmit power of the reader. G_t is the gain of the reader transmit antenna. G_r is the gain of the tag's antenna.d is the distance from tag to reader.

Substituting Eq. (3.45) into Eq. (3.46), the maximum reading distance in the salt mist environment can be calculated as:

$$d = \frac{\lambda}{4\pi}\sqrt{\frac{P_t G_t \tau\left(G_r - \sqrt{\frac{2\sigma}{\omega\mu}}\frac{\eta_1 \eta_2^2 G_r a^2}{(\eta_1+\eta_2)^2 \pi R^2}\right)}{P_{th}}} \tag{3.47}$$

Assuming $y = \frac{\eta_1 \eta_2^2 G_r a^2}{(\eta_1+\eta_2)^2 \pi R^2}\sqrt{\frac{2}{\omega\mu}}$, $z = \sqrt{\frac{P_t G_t \tau}{P_{th}}}$, Eq. (3.47) can be written as:

$$d = \frac{\lambda}{4\pi}\cdot z \cdot \sqrt{G_r - y\sqrt{\sigma}} \tag{3.48}$$

According to the content of Table 3.6, the conductivity increases with the concentration of NaCl solution (x), we assume that there is a nonlinear relationship between the two Physical quantities, and Eq. (3.48) can be written as:

$$d = \frac{\lambda}{4\pi}\cdot z \cdot \sqrt{G_r - y\sqrt{bx^c}} \tag{3.49}$$

We set $b = 7.274$, $c = 0.8405$, $\lambda = C/f = 0.33\text{m}$, $\omega = 2\pi f = 1840\pi \times 10^6 \text{r/s}$, $P_t G_t = 33\text{dBm}$, $G_r = 2\text{dBi}$, $P_{th} = -10\text{dBm}$, $\mu = 4\pi \times 10^{-7}\text{N}\cdot\text{A}^{-2}$, $\eta_2 = 41.8\Omega$, $\eta_1 = 377\Omega$, $R = 0.08\text{m}$, $a = 0.16\text{m}$. When the distance between the tag and surface is about $\lambda/4$, the reading rate is the best. When the distance is $\lambda/2$, the reading rates is worse than that of $\lambda/4$ or $3\lambda/4$ [27]. If the distance of tag to the surface is $\lambda/4$, the transmission coefficient $\tau \approx 1$[35]. Substitute the data into Eq. (3.49), the result is shown in Fig. 6.15. The maximum dynamic reading distance of the RFID decreases as the concentration of the salt mist environment increases.

(2) Dynamic measurement system

We design a dynamic measurement system via a high precision laser sensor to gauge the distance between the sensor and the tag. When different concentrations of salt mist environment experiment boxes are placed in the dynamic system, the maximum reading distance of different concentrations of salt mist environment can be calculated.

The test chamber is designed to simulate different concentrations of salt mist environment. Figure 3.16 shows the salt mist environment experiment box, and Fig. 6.17 shows the experiment box entity diagram.

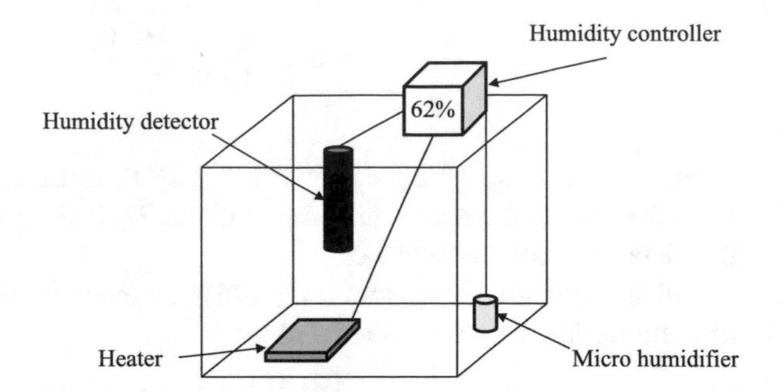

Fig. 3.16 Salt mist environment test chamber

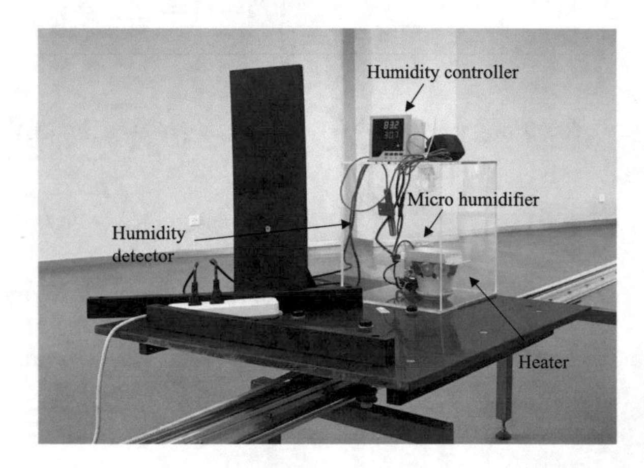

Fig. 3.17 Entity diagram of test chamber

The humidity sensor is suspended in the test chamber. The heater and humidifier, connected to a humidity controller outside the test chamber, are placed on the bottom of the test chamber. We set the experimental humidity of the humidity controller. The humidifier and heater can be humidified or dehumidified as needed. The humidifier is externally connected with different concentration of NaCl solution, and the humidification process is sprayed with NaCl solution inside the test chamber to achieve the salt mist environment required for the experiment. When the hygrometer shows that the humidity reaches 80%, the humidifier stops humidifying, forming an application environment with certain humidity.

In order to simulate the salt mist environment for RFID tag reading experiments, we designed an RFID dynamic measurement system. As shown in Fig. 3.18, the experimental platform consists of the following parts: a control computer, a reader antenna, a laser ranging sensor and other auxiliary equipment. The RFID dynamic measurement system entity diagram is shown in Fig. 3.19.

When the test chamber enters the reading area, the antenna emits electromagnetic waves to activate the tag, then the tag reflects the electromagnetic waves to transmit the information in the tag, and the reader activates the laser ranging radar. Therefore, the reading distance from the antenna to the tag can be calculated, the average of multiple measurements is used for the reading distance of the antenna to tag.

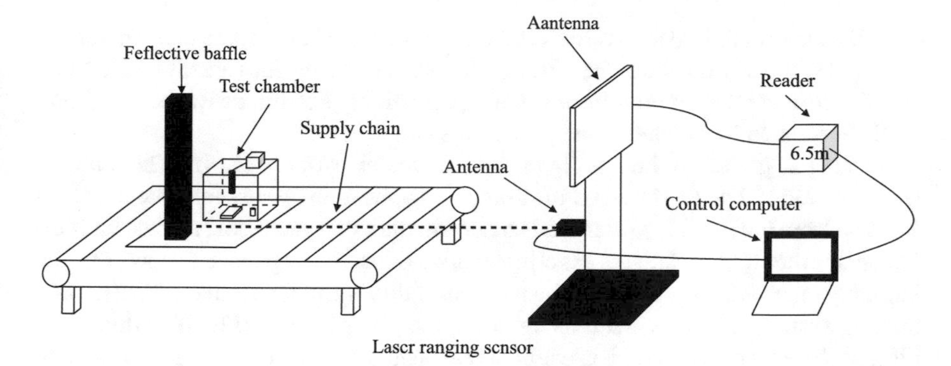

Fig. 3.18 Schematic of the dynamic test system

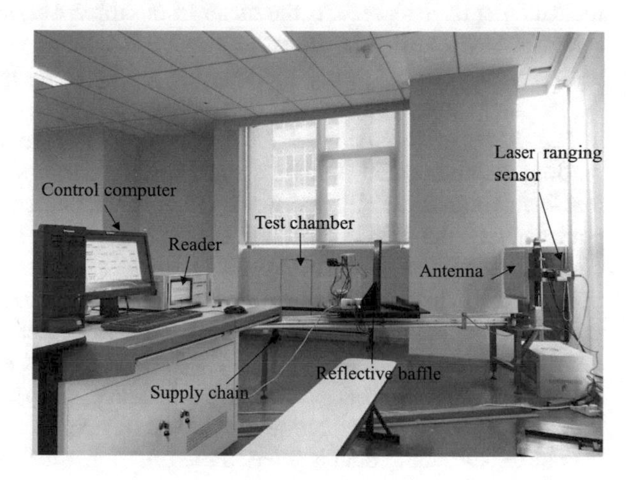

Fig. 3.19 Entity diagram of RFID dynamic measurement system

(3) Experimental results and analysis

The material of the test chamber is polyvinyl chloride, considering the influence of the thickness of the test chamber on the reading performance of the RFID tag. We tested the test chamber with thicknesses of 0.3, 0.5, 1, 2 and 3 mm at room temperature, as shown in the Table 3.7.

Table 3.7 Experimental data of the thickness of test chamber

Thickness of plastic box/mm	Reading distance/m
0.3	5.975
0.5	6.148
1	5.896
2	6.120
3	6.026

In the range of 0.3 ~ 3 mm, all the reading distances are almost the same. Therefore, the test chamber's thickness has little influence on the performance of RFID tag. Small errors can be ignored in practical applications. For the convenience, 3 mm thick plastic box is selected in the proposed system.

The tag is placed in the middle of the test chamber (8 cm ($\lambda/4$)). The ambient temperature is 22.4 °C. Based on the practical application, we assume that the NaCl solution from 0.01 mol/L to 2 mol/L is placed in the experiment box, and the humidifier sprays the solution into the experiment box. When the hygrometer shows that the humidity reaches 80%, the humidifier stops humidifying and starts the RFID dynamic ranging system. The maximum recognition distance of UHF RFID is within 10 m [36], so the experimental data is within a reasonable range. For each situation, three experimental measurements were carried out and the results were expressed as the mean of the three measurements. d_1, d_2, d_3 represent the experimental data of three measurements. $\overline{d}$ represents the mean of the three measurements. Reading range in salt mist environment is shown in Table 3.8. Average reading distance that varies with salt mist environment concentration is shown in Fig. 3.20.

The measurement results have roughly consistency with the simulation results in Fig. 3.20. As the increase of salt mist environment concentration, the dynamic reading distance of RFID tags decrease. Minor errors are caused by errors in the test chamber and the RFID system. When the tag antenna is placed in a conductive environment, the gain pattern and antenna efficiency are reduced due to boundary conditions [37]. As shown in Eq. (3.49), increase in conductivity results in a decrease in the reading distance. Because the tag antenna is placed in a salt mist environment, the electromagnetic waves emitted by the antenna eddy currents will result in a decrease in the electromagnetic energy absorbed and emitted by the tag antenna. Eventually, the feedback electromagnetic wave energy received by the antenna is reduced resulting in a reduction in the read distance.

The experimental data is fitted to the function after trying various functions. The curve matching the power function has the best effect, which is consistent with the result of the formula. The fitting function is shown in Fig. 3.21. The various parameters of the fitting function are as follows:

The fitting function uses the Allometric2 model of the power function, and the equation is as follows:

$$y = a + bx^c \tag{3.50}$$

The numerical ranges of the three parameters are respectively $a = 5.22788 \pm 0.06528$, $b = -2.25495 \pm 0.07286$, $c = 0.60844 \pm 0.02511$. According to statistical theory analysis, the closer the correlation coefficient R_2 is to 1. The smaller the residual sum of squares is, the better the fitting effect of the model is. The correlation coefficient of this fitting function is $R^2 = 0.99659$. The residual sum of squares is 0.05605. Residual sum of squares is 0.05605. The fitting effect of Eq. (3.50) is good.

Then, the absolute error and relative error are calculated by Eqs. (3.51) and (3.52).

$$S = |y_1 - y_2| \tag{3.51}$$

Table 3.8 Reading range in salt mist environment

Concentration/(mol/L)	d_1	d_2	d_3	$\overline{d}$
0	5.544	5.422	5.068	5.345
0.01	5.144	5.057	5.078	5.093
0.1	4.762	4.684	4.453	4.633
0.2	4.502	4.223	4.433	4.386
0.3	4.239	4.028	4.105	4.124
0.4	4.102	3.833	3.981	3.972
0.5	3.706	3.798	3.974	3.826
0.6	3.700	3.532	3.631	3.621
0.7	3.606	3.403	3.482	3.497
0.8	3.102	3.357	3.276	3.245
0.9	3.166	2.951	2.889	3.002
1.0	3.006	2.794	2.912	2.904
Concentration/(mol/L)	d_1	d_2	d_3	$\overline{d}$
1.1	2.621	2.736	2.022	2.793
1.2	2.544	2.688	2.784	2.672
1.3	2.597	2.573	2.354	2.508
1.4	2.476	2.503	2.440	2.473
1.5	2.382	2.126	2.671	2.393
1.6	2.168	2.306	2.393	2.289
1.7	2.007	2.233	2.231	2.157
1.8	1.932	2.048	2.119	2.033
1.9	1.865	2.002	1.839	1.902
2.0	1.736	1.805	1.694	1.745

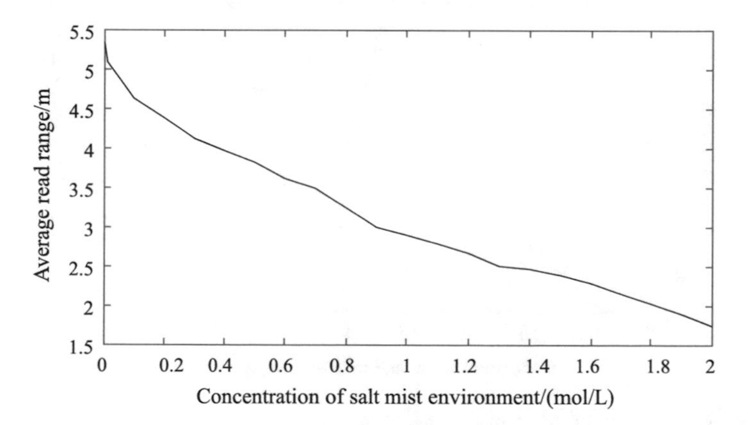

Fig. 3.20 Average reading distance varies with salt mist environment concentration

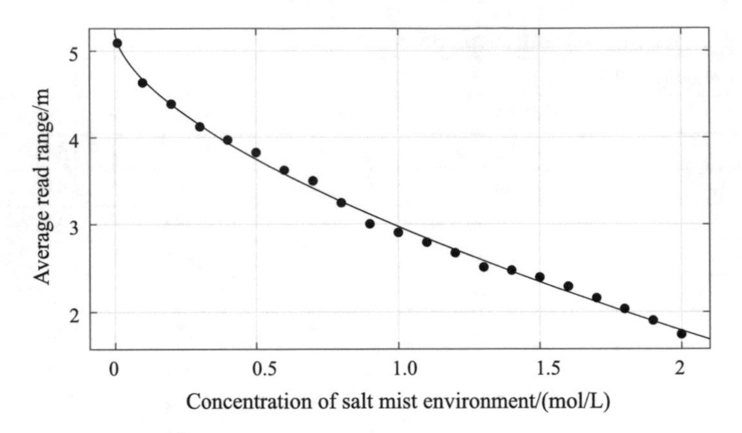

Fig. 3.21 Experimental data fitting function

$$Q = \frac{S}{y_1} \times 100\% \tag{3.52}$$

y_1 is the experimental data.y_2 is the fitted function value. S is the absolute error. Q is the relative error. The relative error is shown in Fig. 3.22.

According to the contents of Fig. 3.22, the absolute error is very small, the relative error is less than 5%, and the maximum relative error is only 3.7%. The fitted function has two advantages. One is that the prediction value of the fitted function is closing to the experimental value, and the absolute error and relative error are very small, which has good practicality. Another is that the fitting function curve is roughly in line with the simulation result curve, which has a good theoretical basis.

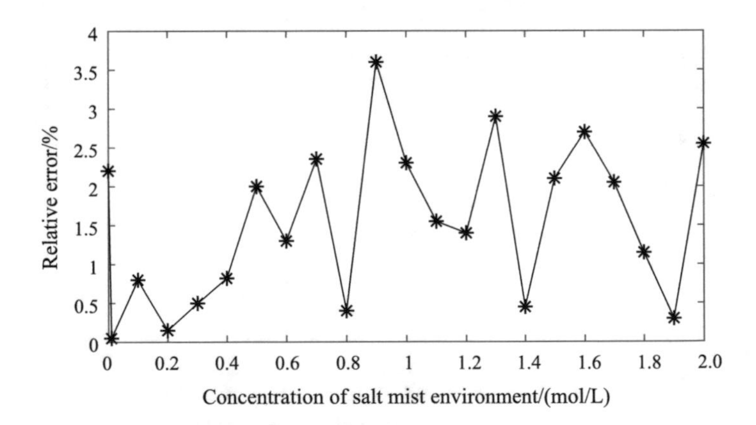

Fig. 3.22 Relative error of different concentrations

3.3 Kinetic Analysis of Physical Anti-Collision—Fisher Information Matrix

In recent years, as one of the core of internet of things (IoT), radio-frequency identification (RFID) has been widely used. With the rapid development of RFID and multi-input multi-output (MIMO) communication, RFID-MIMO system has attracted a wide spread attention in the IOT industry. In RFID, MIMO technology improves reliability of the system by using near field spatial multiplexing and far field spatial diversity to eliminate the interference. Common MIMO system uses several frequencies through multiple channel links while MIMO channel has multiple links operating at the same frequency, which can lengthen RFID reading distance, reduce the bit error rate of the system and improve the reading efficiency of RFID tag and the signal bandwidth doesn't increase [38, 39]. For RFID-MIMO system, Terasaki made experiments on passive MIMO transmission with load modulation. The results indicated that the transmission rate of passive MIMO was up to 2 times higher than that of single-input single-output (SISO) with the same transmission power when the distance between the reader and the tag was 0.5 m. These results also indicate that passive MIMO offers high-speed data transmission even when the distance is doubled [40]. He studied the performance of MIMO RFID backscattering channels, and found out that these properties of the MIMO RFID channel are significantly different from that of other types of cascaded channels such as keyhole and double scattering [41]. In order to get effective biased estimators of the RFID multiple tag system, Yu took the linear and nonlinear functions in the measurement of RFID tags as a paradigm to study the equivalence between the bias-corrected estimators and biased estimators [42]. Aouadi and Belgacem presented an evaluation of MIMO antennas, with analysis of the mutual coupling, correlation coefficient, diversity gain and correlation efficiency. They found that the use of the novel LHM (Left Handed Metamaterial) with multiple bands after fractional removal of substrate was suitable for MIMO antenna system [43]. Xu studied the uplink transmission in MIMO systems with antenna correlation and proposed a very efficient scheme in multi-user uplink MIMO system with distributed channel information [44]. However, few studies have been done about the influence of multiple tags' distribution on the reading performance in RFID-MIMO system. Therefore, this paper focuses on the multiple tags' distribution in RFID-MIMO system.

In the MIMO system, the frequency deviation between transmitter and receiver is mainly caused by the error between the transmitter and the receiver, and the Doppler shift caused by the motion of the mobile station. The problem of multi frequency deviation in MIMO system makes it more complicated to estimate the frequency deviation correctly, and it has become a joint estimation of multi parameter. The multi parameter joint estimation based on ML principle can almost approximate the ideal performance with the increase of signal-to-noise ratio, so it becomes a very practical estimation method [45]. However, the above method has high complexity. Therefore, suboptimal quasi maximum likelihood algorithm is also a common method for the study of multi parameter estimation [46, 47]. Considering the parameter estimation of the MIMO system, Cramer–Rao bound (CRB) determines a lower limit for any unbiased variance of estimator, which means that it is impossible to obtain the unbiased estimator when the variance is less than the lower limit. Meanwhile, the CRB provides a standard of comparing the performance of unbiased estimator. In [48], Wang studied the CRB for joint RSS/DoA-Based Primary-User localization in cognitive radio networks. Bekkerman studied the CRB for estimating parameters of coherently distributed targets (CDT) using MIMO radars [49]. Jagannatham discussed the Cramer–Rao bound based mean-squared error and throughput analysis of superimposed pilots for semi-blind multiple input multiple output wireless channel estimation [50]. In [51], Kalkan calculated the CRB for target position and velocity estimations for widely separated MIMO radar. However, at present, the research on CRB in the MIMO system is mainly focused on the study of the beam and the estimated CRB. There are fewer reports on the RFID-MIMO system, especially the impact of the size of CRB on the reading performance of the RFID- MIMO system. Therefore, on the basis of the above research, this paper introduces the MIMO system into the field of RFID, and constructs a channel model of the RFID-MIMO system. Besides that, in the paper, CRB will be applied to the field of parameter estimation in the RFID-MINO system. The CRB of different tag numbers is simulated by computer to obtain the CRB corresponding to the number of different tags in the case of orthogonal and coherent signals. Finally, by studying the relationship between multi-tag distribution CRB and reading performance, the paper analyzes the effect of multi-tag distribution on reading performance in RFID-MIMO system based on CRB.

3.3.1 Channel Model

Consider a RFID-MIMO system composed of M antennas coordinate of $Z_m = (x_m, y_m)^T (m = 1,\ldots,M)$. Let antenna azimuth angle θ denotes the angle between the tag and the vertical plane of the antenna arrays. The space between two tags is d_a times of wavelength, and the space between two antennas is d_b times of wavelength. The schematic diagram of the structure of RFID-MIMO system is shown in Fig. 3.23.

Let the column vector $s[n]$ represents the baseband signal transmitted by antenna unit when n represents the time. So a echo signal of the target received by the receiving array is given by

$$y[n] = \alpha A(\theta)s[n] + w[n](n = 1, 2, \ldots N) \tag{3.53}$$

where α is the complex amplitude corresponding to the tag, $A(\theta)$ is the corresponding response matrix, $w[n]$ is the noise matrix, and $A(\theta)$ can be defined as

$$A(\theta) = a(\theta)a^{\mathrm{T}}(\theta) \tag{3.54}$$

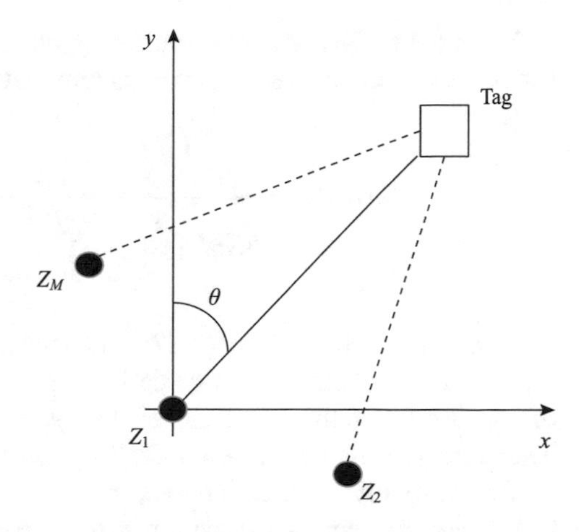

Fig. 3.23 The schematic diagram of the structure of RFID-MIMO system

The coherent matrix of the transmitted signal is given by

$$R_S = \begin{bmatrix} 1 & \beta_{12} & \cdots & \beta_{1M} \\ \beta_{21} & 1 & \cdots & \beta_{2M} \\ \vdots & \vdots & & \vdots \\ \beta_{M1} & \beta_{M2} & \cdots & 1 \end{bmatrix} \tag{3.55}$$

where $a(\theta)$ denotes the steering vector, which is the complex correlation coefficient between the i and jth tag. When the transmit beam of antenna points to the normal direction, the correlation coefficient's phase of transmitted signal is zero, and $\beta_{ij} = \beta_{ji} = \beta$ ($\beta \in [0,1]$). Therefore, when the coherent signal is transmitted, $\beta = 1$. However, when the orthogonal signal is transmitted, $\beta = 0$.

In the white Gaussian noise (WGN) environment, $SNR = N|\alpha|^2/\delta^2$ is signal-to-noise ratio, N is the sampling point, δ^2 is the variance of the sampling signal, and $|\alpha|$ represents the complex amplitude of the received signal. Then the estimation of the CRB for the single tag's space parameter can be expressed as [46]

$$CRB(\theta) = \frac{1}{2SNR(Ma^H(\theta)R_S^{\mathrm{T}}a(\theta) + a^H(\theta)R_S^{\mathrm{T}}a(\theta)\|a(\theta)\|^2)} \tag{3.56}$$

From Eq. (3.56), CRB is inversely proportional to signal-to-noise ratio $N|\alpha|^2/\delta^2$, namely the greater the sampling point, the higher the signal-to-noise ratio, the smaller the CRB, and the better the estimation performance for RFID-MIMO system. The CRB is related to the steering vector of the signal $a(\theta)$ and the number of tags. When the number of tags is given, different CRB can be obtained by changing the waveform of transmitted signal. Therefore, the CRB can be used as a criterion of waveform optimization.

If α is unknown and the transmitted signal is orthogonal to each other, the coherent matrix of the transmitted signal R_s is unit matrix. Hence, the CRB in Eq. (3.56) can be written as

$$CRB(\theta) = \frac{1}{8NM\frac{|\alpha|^2}{\sigma^2}\left(\sum_{k=-(M-1)/2}^{(M-1)/2} k^2\right)(\pi\cos\theta)^2(d_b^2 + d_a^2)} \tag{3.57}$$

It can be shown from Eq. (3.57), when the transmitted signal is orthogonal to each other, CRB is related to the space between two tags, antenna unit and the number of tags. And with the increasing of the space between two tags, the CRB decreases. Therefore, we can obtain better CRB by increasing the space between two tags.

If α is already precisely known, we only need to estimate θ and δ^2. As estimation of δ^2 doesn't affect the estimation of θ, the estimation of the CRB can be expressed as [49]

$$CRB(\theta) = \frac{1}{2SNR(Ma^H(\theta)R_S^{\mathrm{T}}a(\theta) + a^H(\theta)R_S^{\mathrm{T}}a(\theta)\|a(\theta)\|^2)} \tag{3.58}$$

where ||•|| is the norm of a matrix.

The CRB always increase with the estimation of more parameters, because the steering vector considers the center of tag unit as the reference point. When the transmitted signal is orthogonal, the item 3 of the denominator in (3.56) will be equal to 0, the formula (3.56) will be equal to (3.58). Therefore, whether the target range is known, it has no effect on the CRB estimation of RFID-MIMO system which transmits orthogonal signal.

In time-division duplex (TDD) adaptive MIMO system, channel reciprocity is an inherent characteristic of TDD system channels [38]. In TDD system, uplink and downlink channels work in the same frequency. Therefore, on the path that electromagnetic waves transmit, electromagnetic waves which return and go in two directions will through the same reflection, refraction, diffraction and other physical disturbance. Accordingly, it can be assumed that uplink and downlink channels have the same fading characteristics, so we can put the channel state of the uplink channels as channel state of the downlink channels, which is to say that uplink and downlink channels have reciprocity. Let H_u denote the uplink channel state matrix detected on the uplink, and H_d denote the downlink channel state matrix detected on the downlink, so the uplink and downlink channel reciprocity of TDD

$$H_u = H_d^{\mathrm{T}} \tag{3.59}$$

where the superscript T represents the matrix transpose.

Because of the reciprocity theorem, RFID-MIMO system has a plurality of tags and the plurality of antennas which corresponded to a plurality of output and input, and tag along with the antenna works in the same frequency, so it can be shown that the channels of the antenna and the tag also have reciprocity. Therefore, the Eq. (3.56) is introduced into RFID-MIMO system, which represents the number of antennas. For the RFID-MIO system with channel reciprocity, we can use Eq. (3.56) to represent the CRB estimator of antenna azimuth in single-antenna multiple tag system, when M is the number of tags.

3.3.2 Computer Simulation and Analysis

In this section, we imitate the CRB estimator of antenna azimuth in RFID-MIMO system by computer simulating. We can assume that RFID-MIMO system is composed of lined-up tag unit, the space between two tags is half a wavelength, and the centroid is at the origin.

The CRB simulation diagram is shown in Fig. 3.24, where the reader antenna transmits the orthogonal ($\beta = 0$) and coherent ($\beta = 1$) signals, the number of tags are 2, 4, 6, 8, 10, and 100 respectively, and the signal-to-noise ratio is $SNR = 20$ dB It is shown that:

(1) When the antenna azimuth angle θ is close to 90°, the CRB is very large. Regardless of the number of tags or antenna's transmitted signal is orthogonal or coherent, the estimation couldn't be done effectively, which results in a poor reading performance of tags.
(2) When the reader antennas transmit orthogonal signal, with the changing of antenna azimuth angle θ, the CRB is relatively stable, and there is no interference between the reflected signal of two tags. Therefore, the estimation accuracy remains unchanged, and the reading performance of tags is relatively stable.
(3) When the reader antennas transmit coherent signal, with the increase of the antenna azimuth θ, interference between two tags' reflected signal gradually increases which leads to the decrease of the estimation accuracy and CRB increases. Therefore, tags' reading performance declines. And due to the more concentrated interference in a few degrees, the estimation accuracy decreases remarkably. Hence, the peak of the CRB appears at a certain angle, and the peak number increases with the increase of the number of tags. When the number of tags is 8, the distributions of tags and antenna are shown in Fig. 3.25. At this case of Fig. 3.25, the CRB is the largest and the RFID-MIMO system has the worst reading performance.
(4) In addition, the increase of the number of tags decreases the whole system's CRB, which improves the estimation accuracy and the reading performance of tags.

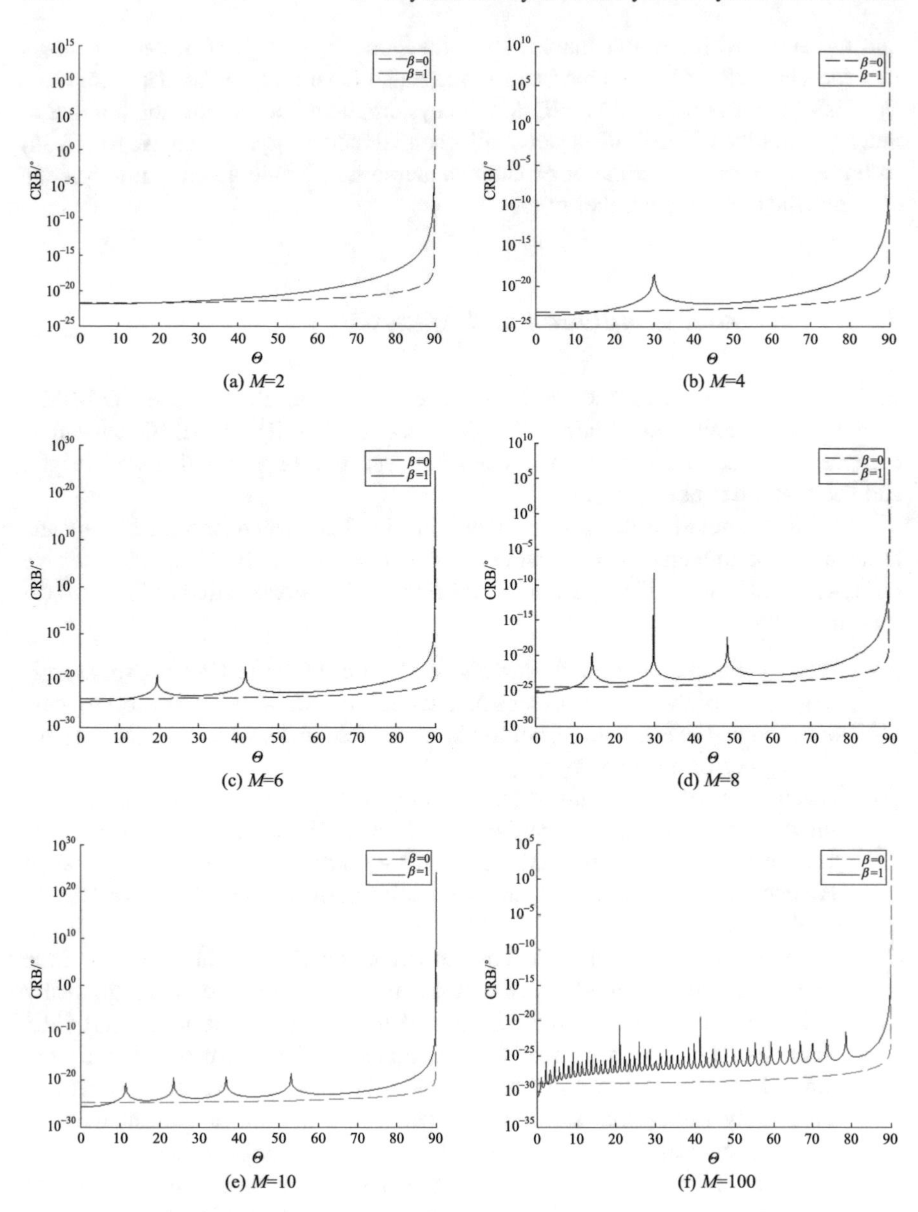

Fig. 3.24 CRB under different number of tags

3.3.3 Antenna Selection Techniques

(1) Optimal antenna selection technique

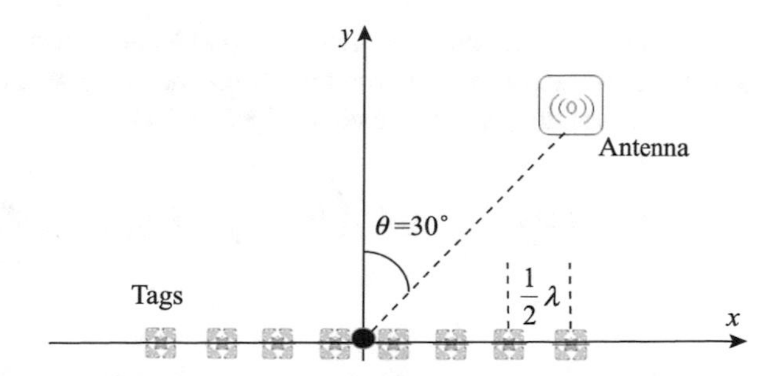

Fig. 3.25 Schematic diagram of tag and antenna position when the tags have the worst reading performance

In RFID-MIMO system, channel capacity improves with the increasing number of RFID tags and antennas. However, the main disadvantage of RFID-MIMO system is the additional high cost for more radio-frequency (RF) modules. Generally, RF modules include low-noise amplifier (LNA), down converter and analog–digital converter (ADC). In order to reduce the cost of multiple RF modules, fewer RF modules are used than the number of reader antennas by applying the antenna selection techniques. A point-to-point distribution of the antenna selection is shown in Fig. 3.26, where only Q RF modules are used to support M_R reader antennas ($Q < M_R$). Therefore, the selected Q RF modules corresponding to Q antennas in M_R reader antennas.

As Q antennas are chosen from M_R reader antennas, Q column of matrix $H \in \mathbb{C}^{M_R \times M}$ represents efficient channel, p_i represents the selected serial number (i, $i = 1, 2, \ldots, Q$) and $H_{\{P_{1,}, P_2, \ldots, P_Q\}} \in \mathbb{C}^{M_T \times Q}$ represents the efficient channel. Let $x \in \mathbb{C}^{Q \times 1}$ represent the space–time code or spatial multiplexing data flow which is mapped to Q selected antennas, the received signal y could be expressed as

$$y = \sqrt{\frac{E_x}{Q}} H_{\{p_{1,}, p_2, \ldots, p_Q\}} x + z \tag{3.60}$$

where E_x is the energy of the transmitted signal and $z \in \mathbb{C}^{M_T \times 1}$ is the additive noise vector. The system capacity in Eq. (3.60) depends on the reader antennas and the corresponding number.

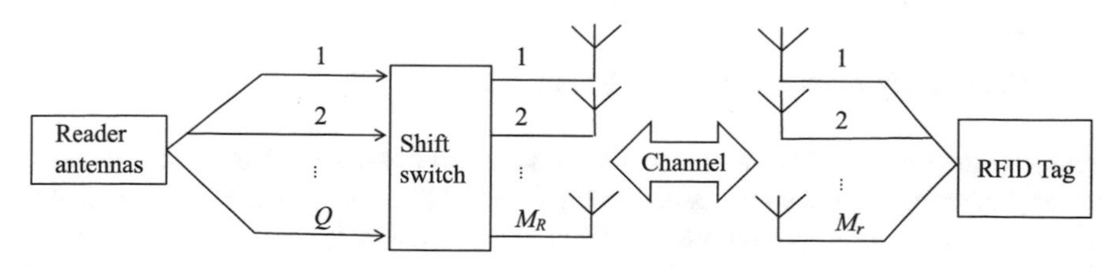

Fig. 3.26 Antenna selection: Q RF modules and MR reader antennas

Q antennas could be chosen from M_R reader antennas to make the channel capacity reach the maximum. When the total transmitted power is P, the channel capacity of Q selected reader antennas can be expressed as

$$C = \max_{R_{xx}, \{p_1, p_2, \ldots, p_Q\}} \log_2 \det\left(I_{M_T} + \frac{E_x}{QN_0} H_{\{p_1, p_2, \ldots, p_Q\}} R_{xx} H^H_{\{p_1, p_2, \ldots, p_Q\}} \right) \text{bps/Hz} \tag{3.61}$$

where R_{xx} is the covariance matrix of $Q \times Q$. If all the selected transmitted antennas have the same power, $R_{xx} = I_Q$. Therefore for $\{p_i\}_{i=1}^{Q}$, the channel capacity can be described as

$$C_{\{p_1, p_2, \ldots, p_Q\}} = \log_2 \det\left(I_{M_T} + \frac{E_x}{QN_0} H_{\{p_1, p_2, \ldots, p_Q\}} H^H_{\{p_1, p_2, \ldots, p_Q\}} \right) \text{bps/Hz} \tag{3.62}$$

The best selection of Q antennas can be realized by calculating Eq. (3.62) with all possible combinations of antennas. In order to maximize system's capacity, antenna with maximum capacity must be selected as

$$\left\{ p_1^{\text{opt}}, p_2^{\text{opt}}, \cdots, p_Q^{\text{opt}} \right\} = \underset{\{p_1, p_2, \cdots, p_Q\} \in A_Q}{\arg\max} C_{\{p_1, p_2, \cdots, p_Q\}} \tag{3.63}$$

where A_Q represents the set formed by all possible combinations of Q selected antennas

$$|A_Q| = \binom{M_R}{Q} \tag{3.64}$$

The simulation of Eq. (3.62) and the curve of channel capacity are shown in Fig. 3.27. The channel capacity of different reader antennas ($M_R = 2, 6, 10, 20$) and tags ($M_T = 2, 6, 10, 20$) as a function of SNR are plotted with different selected antennas Q. It can be seen from the figure that the channel capacity increases linearly with the number of selected antennas. In the case of Fig. 3.27d, when SNR is less than 12 dB, 18 antennas can be selected to ensure the same channel capacity as 20 antennas.

(2) Complexity-reduced antenna selection

As mentioned in the previous subsection, a set of all possible antenna combinations with Q selected antennas is obtained in Eq. (3.64). However, when M_R is very large, all possible antenna combinations in Eq. (3.63) may involve the enormous complexity depending on the total number of available reader antennas. The reading speed and efficiency will be greatly reduced in practical application. In order to reduce its complexity, we may need to resort to the suboptimal method.

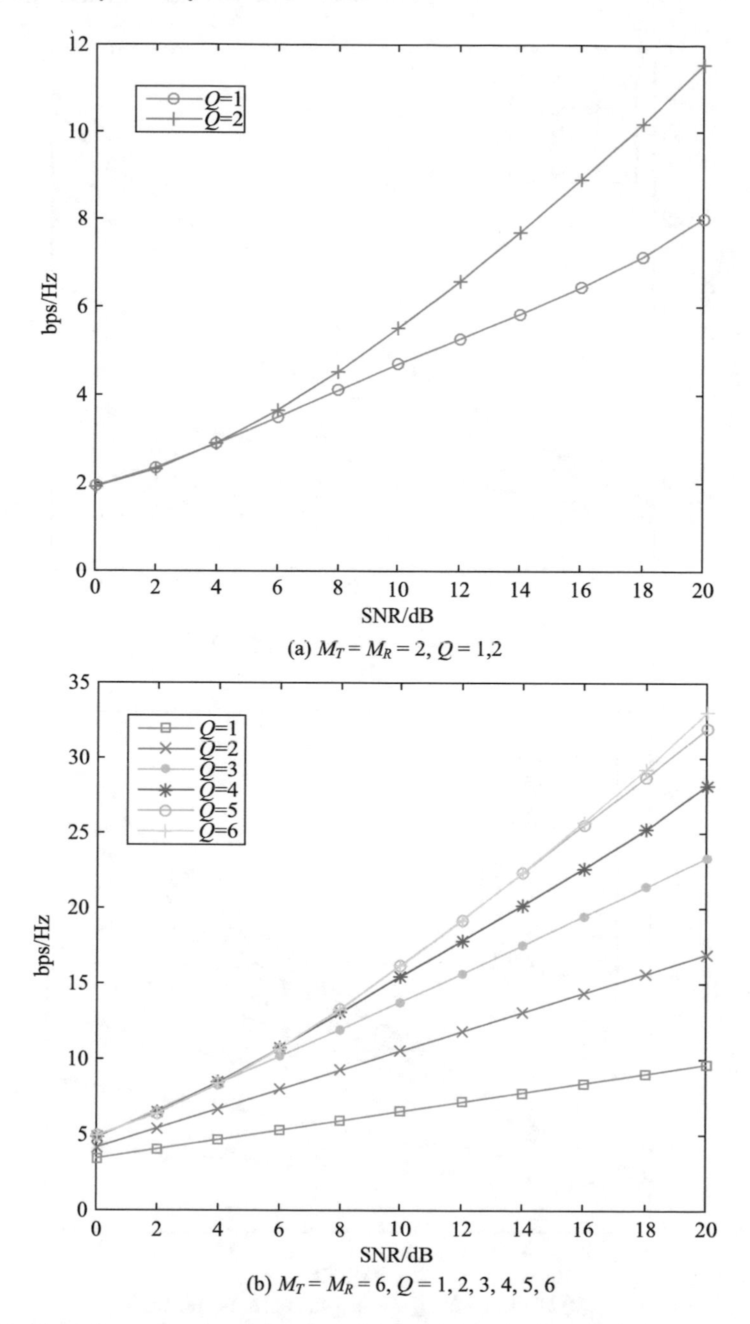

Fig. 3.27 Channel capacity when using the optimal antenna selection technique

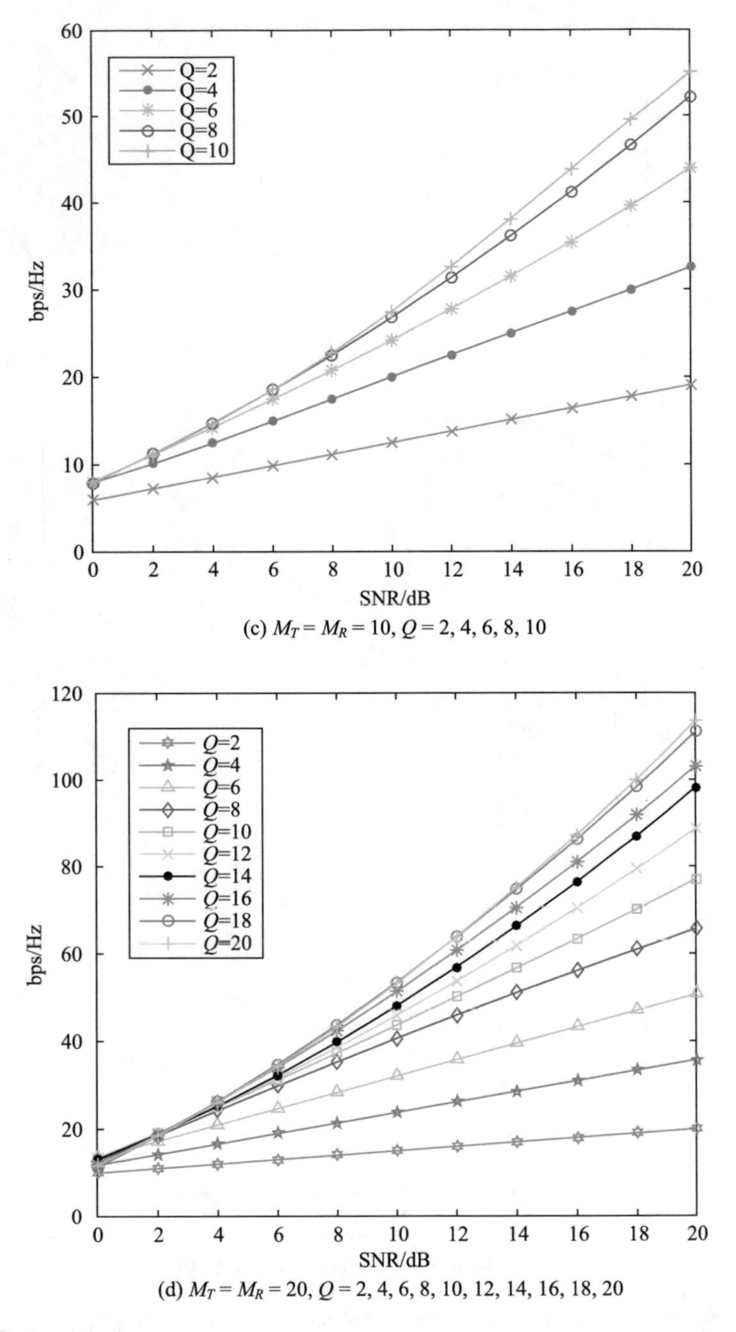

(c) $M_T = M_R = 10$, $Q = 2, 4, 6, 8, 10$

(d) $M_T = M_R = 20$, $Q = 2, 4, 6, 8, 10, 12, 14, 16, 18, 20$

Fig. 3.27 (continued)

Additional antenna can be selected in ascending order of increasing the channel capacity. The first antenna with the highest capacity is selected as

$$p_1^s = \arg\max_{p_1} C_{\{p_1\}} = \arg\max_{p_1} \log_2 \det\left(I_{M_R} + \frac{E_x}{QN_0} H_{\{p_1\}} H_{\{p_1\}}^H \right) \quad (3.65)$$

Given the first selected antenna, the second antenna is selected such that the channel capacity is maximized

$$p_2^s = \arg\max_{p_2 \neq p_1^s} C_{\{p_1^s, p_2\}} = \arg\max_{p_2 \neq p_1^s} \log_2 \det\left(I_{M_R} + \frac{E_x}{QN_0} H_{\{p_1^s, p_2\}} H_{\{p_1^s, p_2\}}^H \right) \quad (3.66)$$

After the nth iteration which provides $\{p_1^s, p_2^s, \ldots p_n^s\}$, the capacity with an additional antenna, antenna v, can be updated as

$$\begin{aligned} C_v &= \log_2 \det\left(I_{M_R} + \tfrac{E_x}{QN_0} H_{\{p_1^s, p_2^s, \ldots p_n^s\}} H_{\{p_1^s, p_2^s, \ldots p_n^s\}}^H \right) \\ &+ \log_2\left(1 + \tfrac{E_x}{QN_0} H_{\{v\}} \left[I_{M_R} + \tfrac{E_x}{QN_0} H_{\{p_1^s, p_2^s, \ldots p_n^s\}} H_{\{p_1^s, p_2^s, \ldots p_n^s\}}^H \right]^{-1} H_{\{v\}}^H \right) \end{aligned} \quad (3.67)$$

The additional $(n + 1)$th antenna is the one that maximizes the channel capacity in Eq. (3.67), that is,

$$\begin{aligned} p_{n+1}^s &= \arg \max_{v \notin \{p_1^s, p_2^s, \ldots p_n^s\}} C_v \\ &= \arg \max_{v \notin \{p_1^s, p_2^s, \ldots p_n^s\}} H_{\{v\}} \left(\frac{QN_0}{E_x} I_{MR} + H_{\{p_1^s, \ldots p_n^s\}} H_{\{p_1^s, \ldots p_n^s\}}^H \right)^{-1} H_{\{v\}}^H \end{aligned} \quad (3.68)$$

This process continues until all Q antennas are selected, (i. e., continue the iteration Eq. (3.68) until $n + 1 = Q$).

Meanwhile, the same process can be implemented by deleting the antenna in descending order of decreasing channel capacity. When $Q = M_R - 1$, the selection method in descending order produces the same antenna index set as the optimal antenna selection method. When $Q = 1$, however, the selection method in ascending order produces the same antenna index as the optimal antenna selection method and achieves better performance than any other selection methods.

The simulation of Eq. (3.67) and the curve of channel capacity are shown in Fig. 3.28. The channel capacity of different reader antennas (M_R = 2, 6, 10, 20) and tags (M_T = 2, 6, 10, 20) as a function of *SNR* are plotted with different selected antennas Q. Comparing the curves in Fig. 6.28 with those in Fig. 3.27, we can see that the suboptimal antenna selection method in Eq. (3.67) achieves almost the same channel capacity as the optimal antenna selection method in Eq. (3.62). Because of the decreasing of the complexity, the calculation speed is greatly improved. However, when the number of reader antennas is very large, the channel capacity

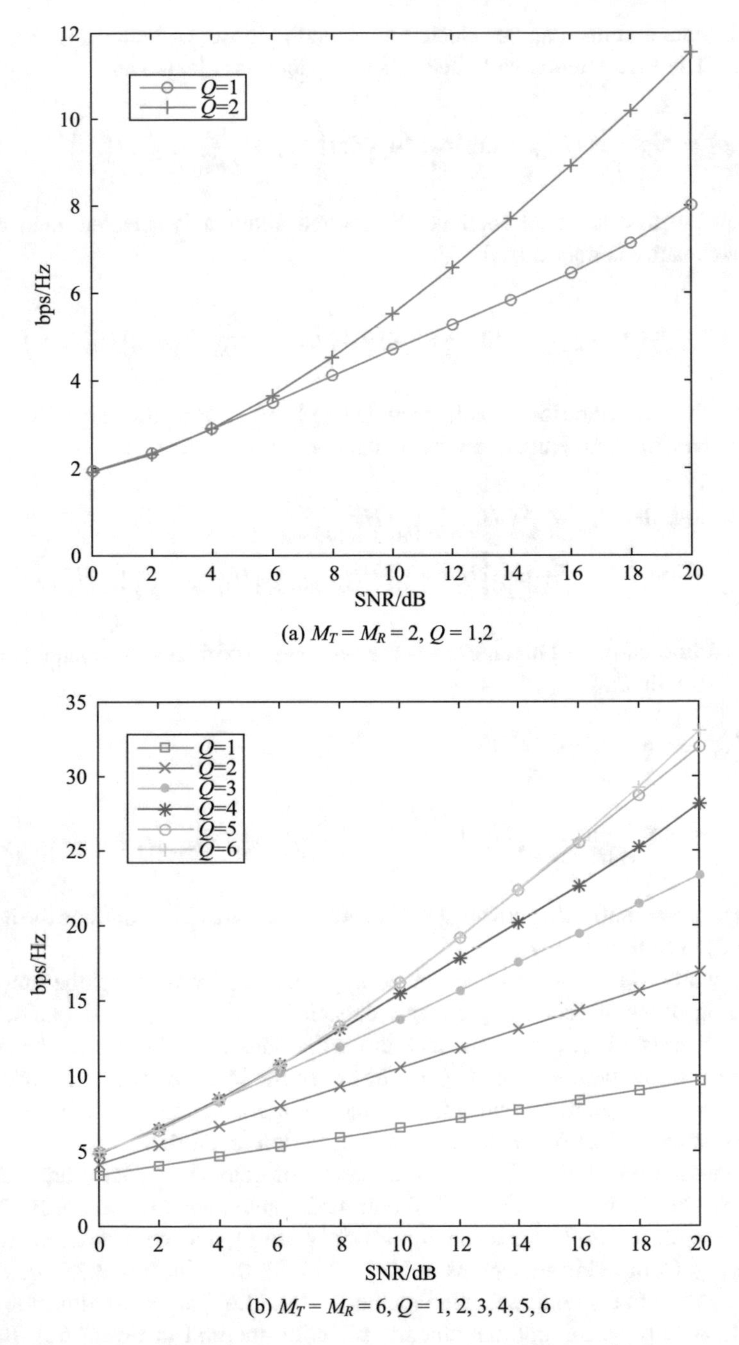

(a) $M_T = M_R = 2$, $Q = 1,2$

(b) $M_T = M_R = 6$, $Q = 1, 2, 3, 4, 5, 6$

Fig. 3.28 Channel capacity when using the complexity-reduced antenna selection

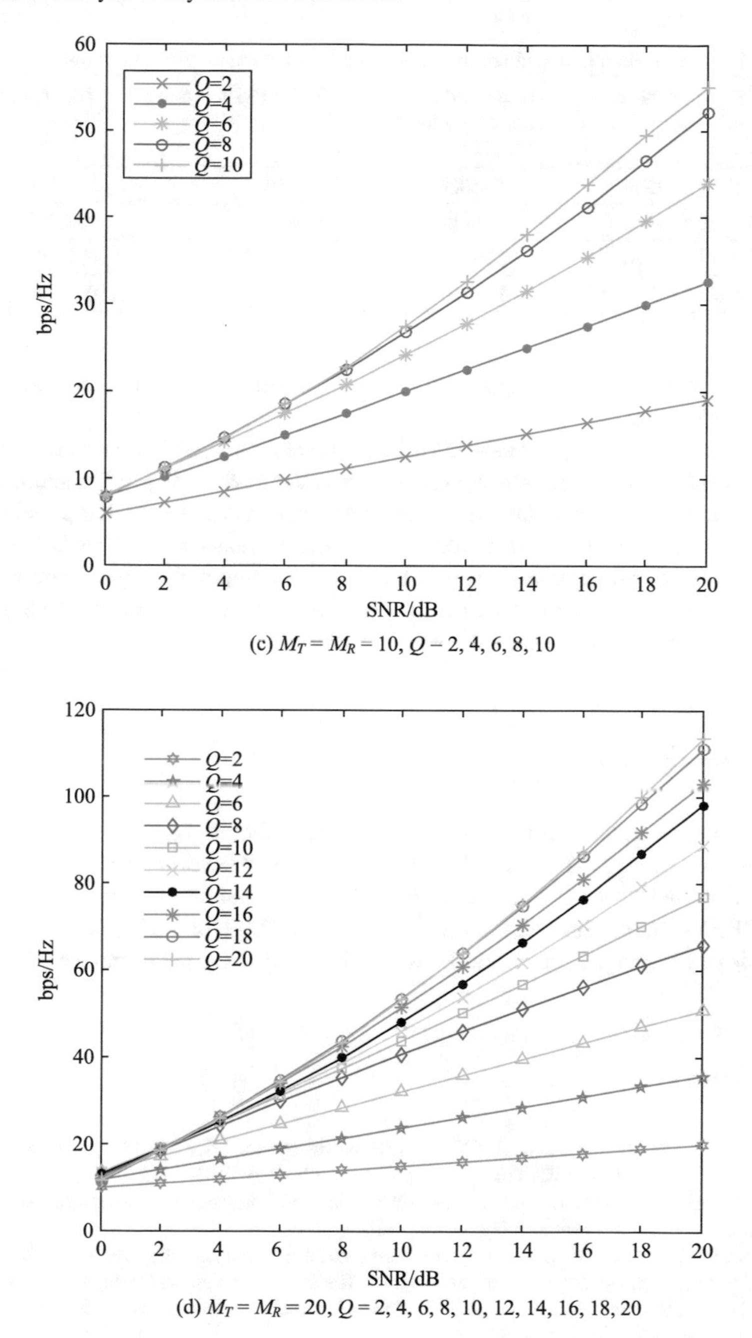

(c) $M_T = M_R = 10$, $Q - 2, 4, 6, 8, 10$

(d) $M_T = M_R = 20$, $Q = 2, 4, 6, 8, 10, 12, 14, 16, 18, 20$

Fig. 3.28 (continued)

Table 3.9 Computation time and relative error of different antenna selection techniques

M	Optimal antenna selection (OAS)/s	Suboptimal antenna selection (SAS)/s	% of OAS's earning	Relative error
6	3.778240	2.450044	64.85%	0.012%
10	9.633543	4.689665	45.68%	0.024%
20	77.049178	18.976021	24.63%	0.065%
50	1021.68677608	124.374018	12.17%	0.214%

of the suboptimal antenna selection will be significantly smaller than the optimal antenna selection.

To compare two different antenna selection techniques, a simulation of computation time and relative error are shown in Table 3.9. Let $M = M_R = M_T$ and $Q = 2$, it can be seen that the percentage of suboptimal antenna selection's computation time reduces and the relative error increases with the increase of M. This is confirmed that the suboptimal antenna selection has a faster computation speed and a higher efficiency. When the number of reader antennas is very large, the relative error of suboptimal antenna selection will be even greater.

3.4 Conclusion

This chapter introduces the influence of different factors on the dynamic reading performance of tags. The control system and RFID detection system are designed to simulate the influence of different factors on the dynamic reading performance of UHF RFID system. This chapter provides a new method for testing UHF RFID tags, which is of great significance for the research of RFID tag performance.

References

1. Rao KVS, Nikitin PV, Lam SF (2005) Antenna design for UHF RFID tags: a review and a practical application. IEEE Trans Antennas Propag 53(12):3870–3876
2. Deleruyelle T, Pannier P, Egels M et al (2010) An RFID tag antenna tolerant to mounting on materials. Antennas Propag Mag 52(4):14–19
3. Kabacik P, Bialkowski ME (1999) The temperature dependence of substrate parameters and their effect on microstrip antenna performance. IEEE Trans Antennas Propag 47(6):1042–1049
4. Yadav RK, Kishor J, Yadava RL (2013) Effects of temperature variations on microstrip antenna. Int J Netw Commun 3(1):21–24
5. Cheng H, Ebadi S, Gong X (2012) A low-profile wireless passive temperature sensor using resonator/antenna integration up to 1000 °C. IEEE Antennas Wirel Propag Lett 11:369–372
6. Babu S, Kumar G (1999) Parametric study and temperature sensitivity of microstrip antennas using an improved linear transmission line model. IEEE Trans Antennas Propag 47(2):221–226

7. Merilampi SL, Virkki J, Ukkonen L et al (2014) Testing the effects of temperature and humidity on printed passive UHF RFID tags on paper substrate. Int J Electron 101(5):711–730
8. Goodrum PM, McLaren MA, Durfee A (2006) The application of active radio frequency identification technology for tool tracking on construction job sites. Autom Constr 15(3):292–302
9. Hahn DW, Ozisik MN (2012) Heat conduction, 3rd edn. Wiley, New York
10. Huleihil M, Andresen B (2006) Convective heat transfer law for an endoreversible engine. J Appl Phys 100(1): 014911.1–014911.4
11. Sheikholeslami M, Ganji DD, Javed MY et al (2015) Effect of thermal radiation on magnetohydrodynamics nanofluid flow and heat transfer by means of two phase model. J Magn Magn Mater 374:36–43
12. Nikitin PV, Rao KVS (2006) Theory and measurement of backscattering from RFID tags. Antennas Propag Mag 48(6):212–218
13. Sha A, Zhang C, Zhou H (2012) The temperature measuring and evaluating methods based on infrared thermal image for asphalt-pavement construction. J Test Eval 40(7):1–7
14. Momma T, Matsunaga M, Mukoyama D, Osaka T (2012) Ac impedance analysis of lithium ion battery under temperature control. J Power Sources 216:304–307
15. Zhang R, Xue A, Gao F (2014) Temperature control of industrial coke furnace using novel state space model predictive control. IEEE Trans Industr Inf 10(4):2084–2092
16. Viswanathan H, Krishnamoorthy R (2001) A frequency offset estimation technique for frequency-selective fading channels. IEEE Commun Lett 5(4):166–168
17. Deacu D (2014) RFID loop tags for merchandise identification onboard ships. Adv Mater Res 1036:969–974
18. Hu Q, Lin LS, Qui YP (2013) Design and implementation of ship through lock in river management system based on rFID. Appl Mech Mater 263:2861–2864
19. Shi X, Tao D, Voss S (2011) RFID technology and its application to port-based container logistics. J Organ Comput Electron Commerce 21:332–347
20. Wu D, Wang D (2011) Experimental Study of UHF RFID performance in liquid environment. IEEE Int Conf Intell Comput Intell Syst 3:511–514
21. Guo J, Wang T, He Y et al (2019) TwinLeak: RFID-based liquid leakage detection in industrial environments. In: IEEE conference on computer communications, pp 883–891
22. Periyasamy M, Dhanasekaran R. Assessment and analysis of performance of 13.56 MHz passive RFID in metal and liquid environment. In: IEEE international conference on communications and signal processing, pp 1122–1125
23. Erdem E, Zeng H, Zhou J et al (2009) Investigation of RFID tag readability for pharmaceutical products at item level. Drug Dev Ind Pharm 35:1312–1324
24. Gutierrez A, Nicolalde F D, Ingle A, et al. High-frequency RFID tag survivability in hash environments. In: IEEE international conference on RFID, pp 58–65
25. Yu Y, Yu X, Zhao Z, Wang D (2017) Influence of temperature on the dynamic reading performance of UHF RFID system: theory and experimentation. J Test Eval 45:1577–1585
26. Merilampi SL, Virkki J, Ukkonen L et al (2014) Testing the effects of temperature and humidity on printed passive UHF RFID tags on paper substrate. Int J Electron 101:711–730
27. Mo L, Zhang H (2007) RFID antenna near the surface of metal. In: Antenna, Propagation and EMC Technologies for Wireless Communications, Microwave, pp 803–806
28. Mo L, Zhang H, Zhou H (2009) Analysis of dipole-like ultra high frequency RFID tags close to metallic surfaces. J Zhejiang Univer-Sci A 10:1217–1222
29. Barge P, Comba L, Gay P. Effect of different chemical compounds concentration in aqueous solution on UHF RFID readability. In: International conference onagricultural engineering-CIGR-AgEng 2016-automation, environment and food safety, CIGR-Ageng, pp 1–5
30. Potyrailo RA, Wortley T, Surman C (2011) Passive multivariable temperature and conductivity RFID sensors for single-use biopharmaceutical manufacturing components. Biotechnol Prog 27(3):875–884
31. Ennasar MA, El Mrabet O, Mohamed K (2019) Design and characterization of a broadband flexible polyimide RFID tag sensor for NaCl and sugar detection. Progress Electromag Res 94:273–283

32. Dean JA. (2003) Lang's handbook of chemistry, 2nd ed. Beijing: Science Press, pp 8.156−8.168
33. Dobkin DM (2008) The RF in RFID: Passive UHF RFID in practice. Elsevier, Amsterdam, Netherland
34. Koski E, Bjorninen T, Ukkonen L (2013) Radiation efficiency measurement method for passive UHF RFID dipole tag antennas. IEEE Trans Antennas Propag 61(8):4026–4035
35. Mo L, Zhang H (2007) RFID antenna near the surface of metal. In: International Symposium on Microwave, Antenna, Propagation and EMC Technologies for Wireless Communications. IEEE, pp 803-806
36. Mo L, Zhang H, Zhou H (2009) Analysis of dipole-like ultra high frequency RFID tags close tometallic surfaces. J Zhejiang Univer-Sci A 10(8):1217–1222
37. Chawla V, Ha DS (2007) An overview of passive RFID. IEEE Commun Mag 45(9):11–17
38. Griffin JD, Durgin GD, Haldi A (2006) RF tag antenna performance on various materials using radio link budgets. IEEE Antennas Wirel Propag Lett 5(1):247–250
39. Marinello JC, Abrao T (2016) Pilot Distribution optimization in multi-cellular large scale MIMO systems. AEU-Int J Electron Commun 70(8):1094–1103
40. Sodagari S, Clancy TC (2013) On singularity attacks in MIMO channels. Trans Emerg Telecommun Technol 26(3):482–490
41. Terasaki K, Honma N (2014) Experimental evaluation of passive MIMO transmission with load modulation for RFID application. IEICE Trans Commun 97(7):1467–1473
42. He C, Chen X, Wang ZJ (2012) On the performance of MIMO RFID backscattering channels. EURASIP J Wirel Commun Netw 1:1–15
43. Yu YS, Yu XL, Zhao ZM et al (2016) Measurement uncertainty limit analysis of biased estimators in RFID multiple tags system. IET Sci Meas Technol 10(5):449–455
44. Aouadi B, Tahar JB (2015) Requirements Analysis of dual band MIMO antenna. Wireless Pers Commun 82(1):35–45
45. Xu C, Wang P, Zhang Z (2015) Transmitter design for uplink MIMO systems with antenna correlation. IEEE Trans Wirel Commun 14(4):1772–1784
46. Jiang Z, Wang H, Ding Z (2013) A bayesian algorithm for joint symbol timing synchronization and channel estimation in two-way relay networks. IEEE Trans Commun 61(10):4271–4283
47. Masmoudi A, Bellili F, Affes S et al (2017) Maximum likelihood time delay estimation from single-and multi-carrier DSSS multipath MIMO transmissions for future 5G networks. IEEE Trans Wireless Commun 16(8):4851–4865
48. Simeunović DM (2018) Review of the quasi maximum likelihood estimator for polynomial phase signals. Digital Signal Process 72:59–74
49. Wang J, Chen J, Cabric D (2013) Cramer-Rao bounds for joint RSS/Doa-based primary-user localization in cognitive radio networks. IEEE Trans Wireless Commun 12(3):1363–1375
50. Tabrikian BJ (2006) Target detection and localization using MIMO radars and sonars. IEEE Trans Signal Process 54(10):3873–3883
51. Jagannatham AK, Rao BD (2014) Cramer-Rao bound based mean-squared error and throughput analysis of superim-posed pilots for semi-blind multiple-input multiple-output wireless channel estimation. Int J Commun Syst 27(10):1393–1415
52. Kalkan Y (2013) Cramer-Rao bounds for target position and velocity estimations for widely separated MIMO radar. Radio Eng 22(4):1156–1161

Chapter 4
Image Theory of RFID System Physical Anti-Collision

In order to collect multi-tag data, this study designed the flat RFID system in the previous article and designed a three-dimensional system based on this. Chapter 3 deblurred the motion blur caused by relative motion in the process of acquiring RFID images. This chapter is based on acquiring the coordinates of the RFID tag group using the system and image matching technology after acquiring a clear image of the RFID tag. This chapter puts forward the research of RFID tag positioning method based on three-dimensional space multi-tag positioning system, which solves the problem of accurate identification of RFID multi-tag spatial distribution in a complex environment, and provides an important reference for improving the research of RFID tag reading performance.

4.1 Tag Distribution Based on Image Matching

In the last chapter, we solved the test method of tag group sensitivity and the influence of different tag distributions on tag group sensitivity. When it is necessary to analyze the performance of multiple tag groups, we use the means of image acquisition and processing to confirm the tag position information and complete the evaluation of tag group performance. In the third chapter of the RFID tag group dynamic test system, the key tag features of the image are extracted by image feature matching method for different positions, and the key tags in the complex environment are effectively matched and located. This chapter verifies the feasibility of the image feature matching method for tag cloud performance evaluation. In this chapter, we propose an improved image matching algorithm, which combines the minimum feature value algorithm to extract feature points and the direction gradient histogram to calculate the description vector. The algorithm is optimized for RFID multi-tag identification and allocation in actual scenarios. The traditional SURF algorithm has the problems of low matching accuracy and high multi-tag matching complexity. This

X. Yu et al., *Physical Anti-Collision in RFID Systems*,
https://doi.org/10.1007/978-981-16-0835-3_4

algorithm has high matching accuracy and real-time performance, and it provides an effective method for RFID multi-tag real-time fast matching.

4.1.1 Algorithm Design

(1) Algorithm flowchart

The image matching process of the improved algorithm proposed in this paper is shown in Fig. 4.1. It is mainly divided into four stages: reading the picture and graying, corner detection by minimum eigenvalue algorithm, calculation of description vector by HOG algorithm, and creation of matching pairs.

(2) Algorithm principle

After reading the picture and graying it, the algorithm first creates a$^{n \times n}$ window, quotes the first-order partial derivative, and then calculates the gray change in the window before and after the sliding when the local small window of the image is moved. The description is as Eq. (4.1):

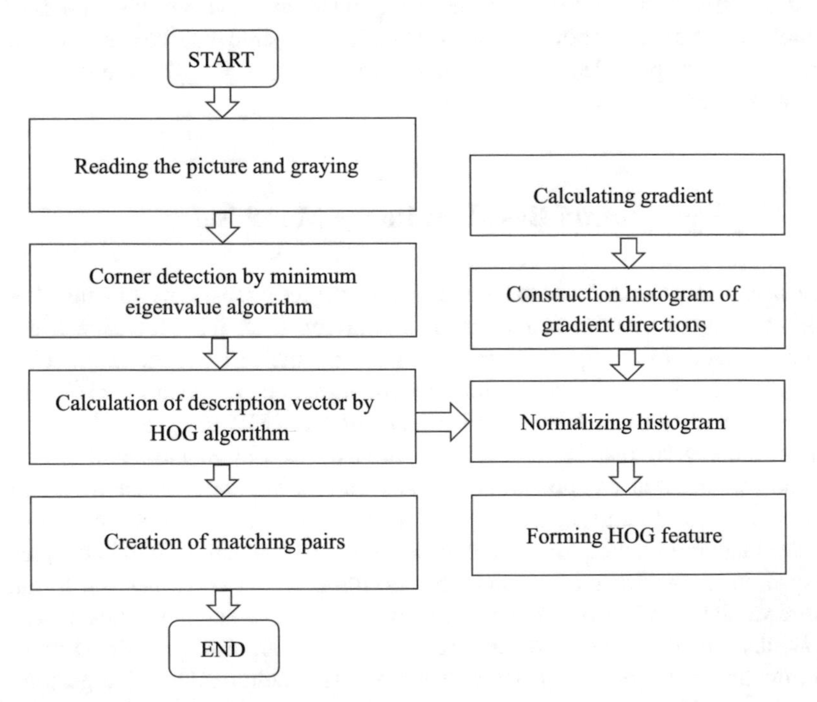

Fig. 4.1 Algorithm flow chart

$$E(u, v) = \sum_{x,y} w(x, y)\big[I(x+u, y+v) - I(x, y)\big]^2 \tag{4.1}$$

Among them, $w(x, y)$ is the window function and $I(x, y)$ is the gray value. In order to do corner detection, the algorithm maximizes this function $E(u, v)$. Expanding Eq. (4.1) by Taylor and omitting higher order terms, the final equation is:

$$E(u, v) \approx \begin{bmatrix} u\ v \end{bmatrix} M \begin{bmatrix} u \\ v \end{bmatrix} \tag{4.2}$$

Among them,

$$M = \sum_{x,y} w(x, y) \begin{bmatrix} I_x I_x & I_x I_y \\ I_x I_y & I_y I_y \end{bmatrix} \tag{4.3}$$

Among them, I_x and I_y are the derivatives of the image in the x and y directions, respectively.

The algorithm defines a score function, which determines whether a window contains a corner as Eq. (4.4):

$$R = \min(\lambda_1, \lambda_2) \tag{4.4}$$

In Eq. (4.4), λ_1 and λ_2 are the characteristic values of M. If the value of formula (4.4) is greater than the minimum threshold value, it is regarded as an angle.

After extracting the feature points, the algorithm calculates the gradient of the abscissa and ordinate directions of the image and calculates the gradient direction value of each pixel position accordingly. The gradient of the pixels in the image is:

$$\begin{aligned} G_x(x, y) &= H(x+1, y) - H(x-1, y) \\ G_y(x, y) &= H(x, y+1) - H(x, y-1) \end{aligned} \tag{4.5}$$

In Eq. (4.5), $G_x(x, y)$ represents the horizontal gradient at pixel (x, y) in the input image, $G_y(x, y)$ represents the vertical gradient, and $H(x, y)$ represents its pixel value. The gradient magnitude and gradient direction at pixel (x, y) are:

$$G(x, y) = \sqrt{G_x(x, y)^2 + G_y(x, y)^2} \tag{4.6}$$

$$\alpha(x, y) = \tan^{-1}\left(\frac{G_y(x, y)}{G_x(x, y)}\right) \tag{4.7}$$

With the above calculation formula, the algorithm divides the image into several image blocks, and each block is divided into several cells. The gradient histogram is constructed in the unit of cell to form 2D vector, the gradient histogram is normalized within the block, and the vectors corresponding to all blocks are arranged in spatial order to form the final feature descriptor.

Finally, when establishing a match, the algorithm determines the matching degree by calculating the Euclidean distance between two feature points. The shorter the distance, the better the match.

4.1.2 Experiment

In order to verify the accuracy of the improved algorithm, the simulation experiment was implemented on a computer with an Intel(R) Core(TM) i5-9300H CPU2.4 GHz, Window 10 operating system, and MATLABR2018a environment. First, the images of this experiment are collected in the RFID dynamic test system [1]. The system is mainly composed of reader antenna, laser rangefinder, CCD camera, drive motor, RFID reader, reflector, guide rail, optical lifting platform, data display, and computer. This experiment analyzes and compares the algorithm in this paper with the SURF algorithm in terms of operating speed and matching accuracy. In order to verify the robustness of the algorithm in this paper, the images selected in the experiment are square RFID multi-tag image, rectangular RFID tag image, and circular RFID tag image.

The square RFID multi-tag image matching results are shown in Figs. 4.2, 4.3,

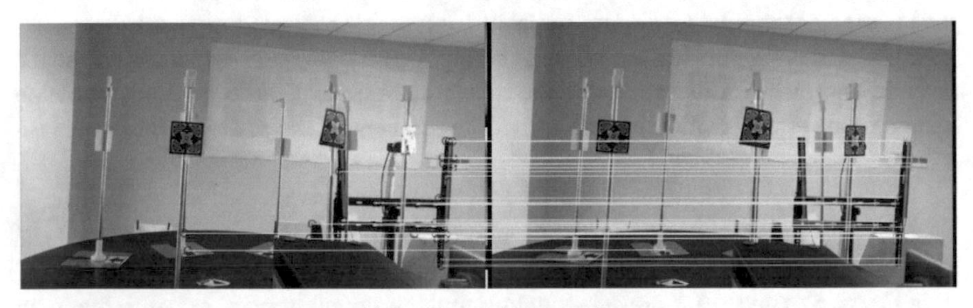

(a) SURF matching results

(b) Matching results in this paper

Fig. 4.2 Square RFID multi-tag image matching results in the first group

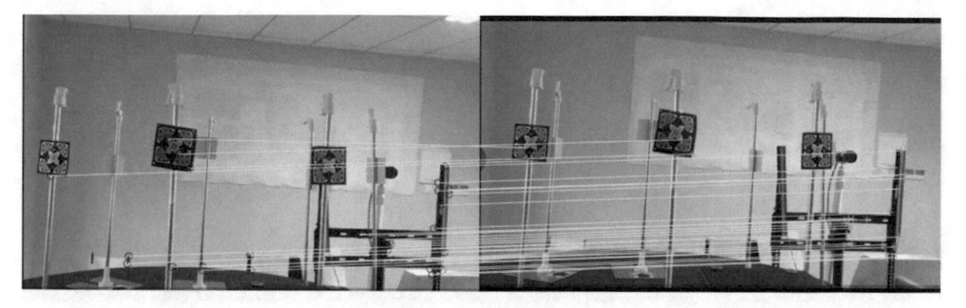

(a) SURF matching results

(b) Matching results in this paper

Fig. 4.3 Square RFID multi-tag image matching results in the second group

and 4.4. The image matching results of the second type of rectangular RFID tags are shown in Figs. 4.5 and 4.6. The image matching results of the third round RFID tag are shown in Fig. 4.7.

For square RFID multi-tag images, the running time, the number of feature points, the number of successful matches, and the matching rate of the two algorithms are shown in Table 4.1. In Table 4.1, the matching rate is shown in Eq. (4.8):

$$A = \frac{N}{F} \times 100\% \tag{4.8}$$

In Eq. (4.8), A represents the matching rate, N represents the number of matches, and F represents the number of feature points.

It can be seen from Table 4.1 that the running speed of the algorithm in this paper is 54.8% higher than that of the SURF algorithm. The average matching rate of this algorithm is 92.0%, which is higher than the 43.9% of the SURF algorithm.

For different other label images, the running time, the number of feature points, the number of successful matches, and the matching rate extracted by the two algorithms are shown in Tables 4.2 and 4.3. It can be seen from Tables 4.2 and 4.3 that for rectangular and circular RFID tag images, the running speed of the algorithm in this

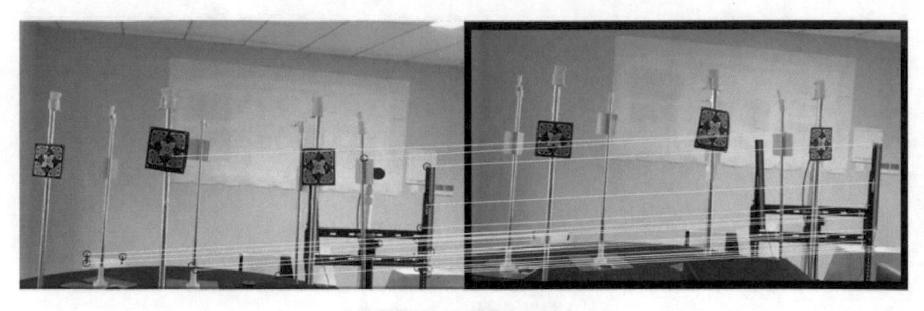

(a) SURF matching results

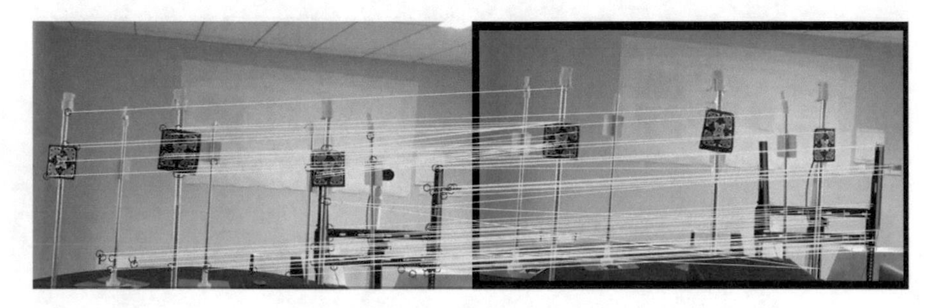

(b) Matching results in this paper

Fig. 4.4 Square RFID multi-tag image matching results in the third group

paper is 57.3% higher than that of the SURF algorithm. The average matching rate of this algorithm is 82.8%, which is higher than the 43.3% of the SURF algorithm.

In order to verify the time and accuracy of the image registration method of this algorithm, Table 4.4 shows the comparison results of the three algorithms. For specific data, see https://weibo.com/7360365656/profile?topnav=1&wvr=6&is_all=1.

Synthesizing Tables 4.1, 4.2, 4.3, and 4.4, it shows that compared with the SURF algorithm and literature algorithm [2], the algorithm in this paper has a better matching effect and takes less time than others.

Aiming at the feature point pairing of RFID tags, this section introduces a fast RFID image matching algorithm for multi-tag identification and distribution optimization. This algorithm is an accurate and real-time algorithm and provides an effective way for multi-tag real-time fast matching.

4.2 Image Processing in Multi-tag Movement

In order to collect multi-tag data, this study designed a planar positioning system in the previous article and designed a three-dimensional positioning system based on the system. In the early stage of design, this paper chose the reader to collect the

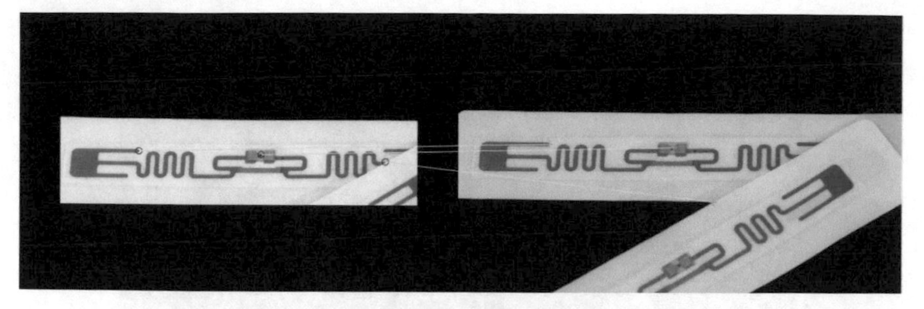

(a) SURF matching results

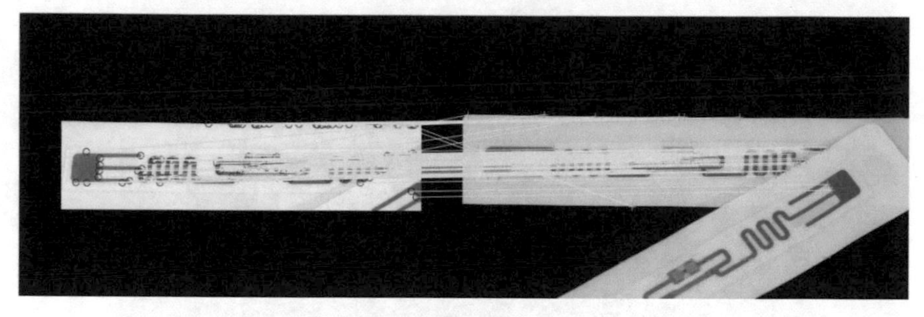

(b) Matching results in this paper

Fig. 4.5 Rectangular RFID tag image matching results in the first group

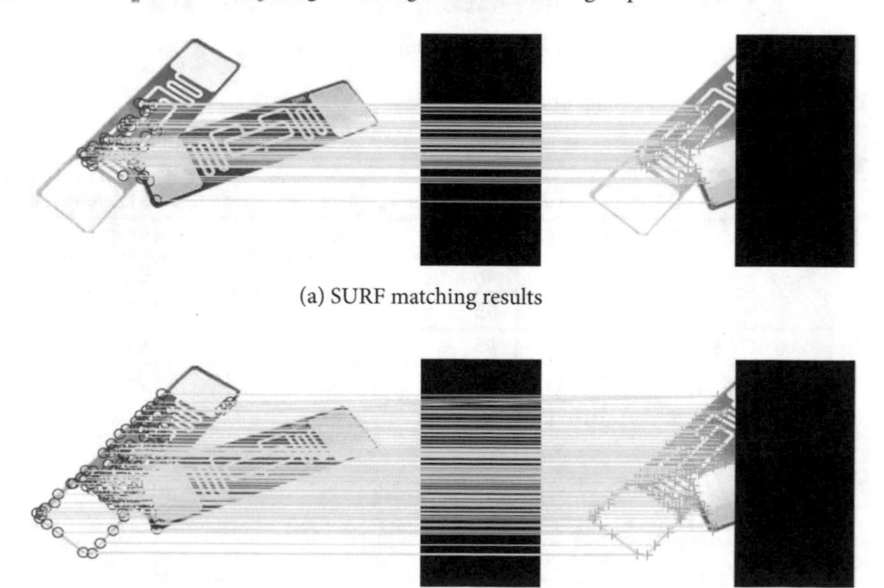

(a) SURF matching results

(b) Matching results in this paper

Fig. 4.6 Rectangular RFID tag image matching results in the second group

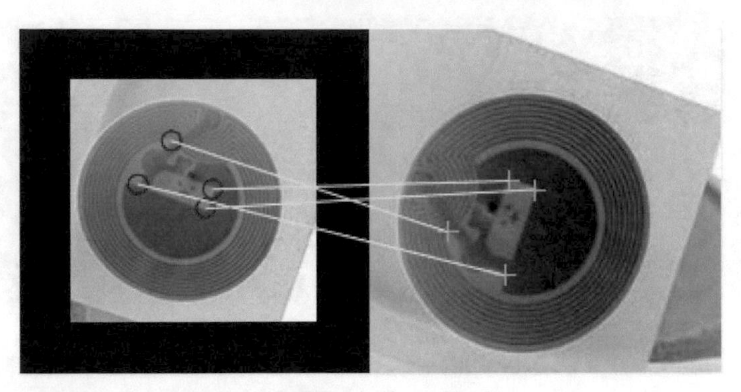

(a) SURF matching results

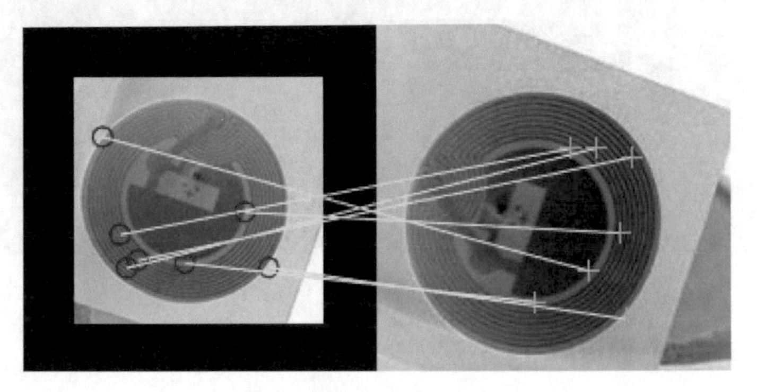

(b) Matching results in this paper

Fig. 4.7 Image matching result of round RFID tag

Table 4.1 Image test results of square RFID tags

	Figure	Time/s	Feature points	Number of matches	Matching rate/%
SURF	3	2.032	82	32	39.0
	4	1.967	48	33	68.7
	5	1.962	96	23	23.9
Improved SURF	3	0.882	93	81	87.1
	4	0.884	105	100	95.2
	5	0.931	98	92	93.8

Table 4.2 Image test results of rectangular RFID tags

	Figure	Time/s	Feature points	Number of matches	Matching rate/%
SURF	6	2.082	52	3	5.76
	7	2.359	95	42	44.2
Improved SURF	6	0.674	85	63	74.1
	7	1.211	121	105	86.7

Table 4.3 Image test results of round RFID tags

	Figure	Time/s	Feature points	Number of matches	Matching rate/%
SURF	8	1.760	5	4	80.0
Improved SURF	8	0.764	8	7	87.5

Table 4.4 Comparison of tag image test results of different algorithms

	Average time/s	Average matching rate/%
SURF	2.027	43.6
Literature [2] algorithm	1.183	80.2
Improved SURF	1.056	87.4

position data of the tag. However, in the actual simulation process, relative motion between tag and camera would affect the position data collected by the reader, so that the system cannot accurately obtain the position of the tag, and then affect the coordinate distribution of the subsequent acquisition of RFID multi-tag. In order to ensure the smooth progress of the follow-up research, the tag position data must be collected by a method that can remove the generated motion blur. Based on the system designed in the previous article, regarding the possibility of multiple distribution states of tags, this paper proposed an algorithm that can meet the needs of use and provides an effective data guarantee for obtaining the three-dimensional coordinates of RFID tags.

4.2.1 Image Deblur Theory

When shooting a moving target, due to the relative motion between the subject and the camera, the acquired image is degraded, and such a blurred image is a motion-blurred image. In many practical applications, motion blur needs to be removed. Under normal circumstances, the process of motion degradation can be modeled as a two-dimensional linear displacement invariant process. In this process, the blurred image $g(x, y)$ can be expressed as the convolution of the original image $f(x, y)$ points and the diffusion function (PSF) [87]; therefore, the restoration process from the blurred image is actually the operation of deconvolution. The degradation model is shown in Fig. 4.8.

The mathematical expression of the model is:

$$g(x, y) = f(x, y) * h(x, y) + \eta(x, y) \tag{4.9}$$

In Eq. (4.9), * represents a two-dimensional linear convolution, $\eta(x, y)$ is additive noise, and $h(x, y)$ represents a point spread function. The expression of the model in

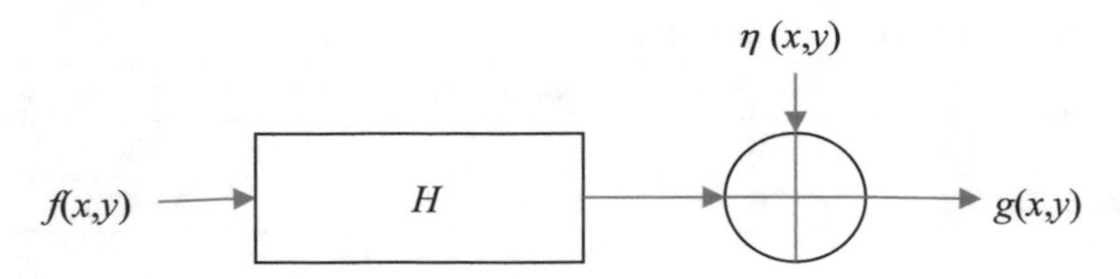

Fig. 4.8 Image degradation model

the frequency domain is:

$$G(u, v) = F(u, v)H(u, v) + N(u, v) \tag{4.10}$$

where $G(u, v)$, $F(u, v)$, $H(u, v)$, and $N(u, v)$ are the Fourier transform of the blurred image, original image, PSF, and noise, respectively.

If a motion blur problem has a known point expansion function, then the problem can be recovered in various ways, for example, Wiener filtering, inverse filtering, Fourier wavelet deconvolution, wavelet-based maximum value algorithm, Wiener filtering with optimal window, etc.

The point spread function of motion blur can be described as

$$h(x, y) = \begin{cases} 1/L \; \sqrt{x^2 + y^2} \leq L/2, y/x = \tan\theta \\ 0 \qquad\qquad \text{other situations} \end{cases} \tag{4.11}$$

In Eq. (4.11), L represents the blurred length and θ represents the blurred angle. For motion blur problems, blur parameters directly affect the degree of blur, such as blurred angle and blurred length. The performance of fuzzy recovery depends on the accuracy of PSF parameter estimation [88, 89], so it is necessary to accurately estimate the length and angle of blur from the given motion blur function. The deblurring process is shown in Fig. 4.9.

Figure 4.9 shows the effect of PSF on the image in the frequency domain. Among them, the abscissa represents the distribution of the frequency spectrum of the picture, the ordinate represents the logarithm of the grayscale of the picture, the curve $\log|U|$

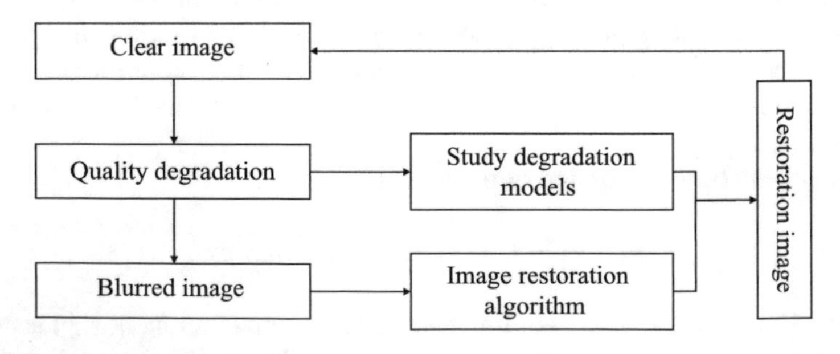

Fig. 4.9 Deblurring process

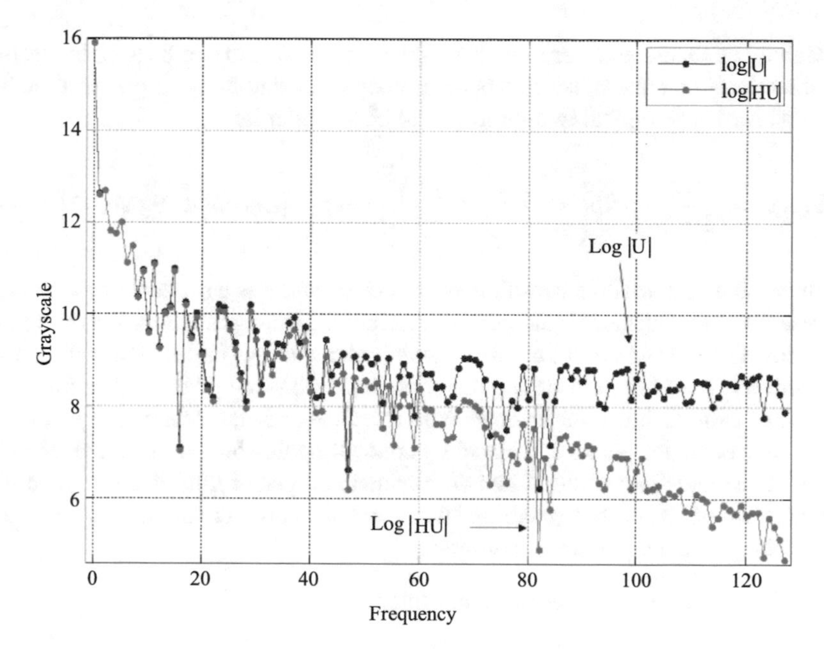

Fig. 4.10 Effect of PSF on images in the frequency domain

represents the distribution of the ideal picture in this coordinate, and the curve $\log|HU|$ represents the distribution of the image after it interacts with the blur kernel (PSF). This means that the effect of PSF on pictures has obvious characteristics in the frequency domain.

Different estimation methods are needed to estimate the blurred length and blurred angle. The blurred angle can be estimated by using the Gabor filter to filter the blurred image in the frequency domain, and the blurred angle can be estimated according to its response. The fuzzy length can be used to train the neural network to estimate the ancient fuzzy length. After obtaining the estimated values of blurred angle and blurred length, the fuzzy point spread function can be constructed. Finally, Wiener filtering can be used to restore the blurred image according to the point spread function.

4.2.2 Estimation of Image Blurred Angle Based on Gabor Filter

(1) Gabor filter basics

The motion blurred image can be found according to the spectrogram to show that the motion blurred image has obvious directional characteristics in the frequency domain.

Gabor is a Gaussian filter modulated by a sine wave and can be used to discover the direction in a pattern, such as pattern recognition and image segmentation. The function form of a typical two-dimensional Gabor filter is:

$$G(x, y) = \frac{1}{2\pi\sigma_x\sigma_y} \exp\left[-\frac{1}{2}\left(\frac{x^2}{\sigma_x^2} + \frac{y^2}{\sigma_y^2}\right)\right] \cdot \exp\left[-j\omega(x\cos\varphi + y\sin\varphi)\right] \quad (4.12)$$

In the spatial domain, a two-dimensional Gabor filter is a product of a sine plane wave and a Gaussian kernel function. The former is a tuning function, and the latter is a window function where σ_x, σ_y are the standard deviations in the x and y directions, respectively. φ and ω indicate the direction and frequency of the Gabor filter. The response of the Gabor filter changes with the change of the orientation parameter, so the blurred angle can be calculated by controlling this direction parameter of the two-dimensional Gabor filter. The two-dimensional Gabor filter is convoluted with the spectrum of the blurred image, and the response in different directions is obtained by keeping the other parameters different.

(2) Obtain blurred angle using Gabor filter

The orientation of the lines in the spectrum of the blurred image can directly affect the judgment of the blurred angle. For line detection algorithms such as Radon transform and Hough transform, it can also be used to detect the direction of the line. Hough transform needs a threshold value to determine the point on any straight line. Different thresholds are required for different images, and any small error in the threshold may cause a large change in the estimation of the blurred angle [88]. One of the motion-blurred images and its spectrum obtained from the experiment are shown in Fig. 4.11.

The Gabor filter to determine the blurred angle can effectively reduce the impact of this problem. Gabor filter response depends on the frequency and direction of the input image. The Gabor filter template is shown in Fig. 4.12.

By detecting the motion direction of the blur graph through Gabor, the blurred angle θ corresponding to the angle φ with the highest response value can be obtained. The maximum response of the filter is calculated using the L_2 norm. This method keeps other parameters unchanged and only changes the direction of the filter. The filter φ with different orientations is convoluted with the Fourier transform of the blurred image. So for each φ, the L_2 norm of the matrix generated by convolution must be calculated. The largest L_2 norm corresponds to the blurred angle. Its concrete steps are:

(1) Calculate the spectrum of the blurred image.
(2) Use the logarithm of the blurred image spectrum $I = \log(G(x, y))$ as the input of the Gabor filter.
(3) Convolute of Gabor filters with different angles φ and I to get the response of each angle $R(\varphi)$.
(4) For each angle φ, calculate the corresponding L_2 norm [90].

(a) Tag image with blurred 20 and angle 45°

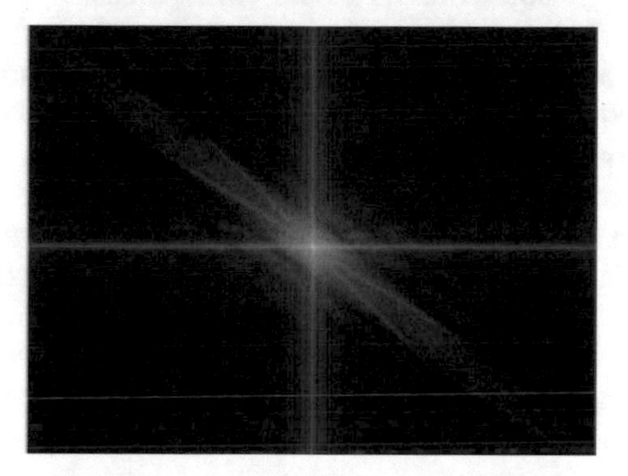

(b) Spectrogram of tag blurred image (a)

Fig. 4.11 Motion blurred image and its spectrum

(a) 30℃ Gabor filter template

(b) 45℃ Gabor filter template

Fig. 4.12 Gabor filter template

4.2.3 Estimation of Image Blurred Length Based on Generalized Regression Neural Network (GRNN)

(1) Generalized regression neural network structure

The second parameter of motion blur is the blurred length. It describes how much an object or camera has moved during the exposure time. In order to predict the blurred length of a specific blurred image, the sum of Fourier coefficients (Sum of the Magnitude of Fourier Coefficients, SUMFC) of the corresponding blurred image can be used as its input. The Fourier feature of the image is one of the simplest features in the frequency domain, and it is easy to determine using the FFT algorithm. According to existing studies, there is a nonlinear relationship between SUMFC and fuzzy length [91]. This nonlinear relationship can be estimated by GRNN, that is, the generalized regression neural network is selected for estimation.

In the early 1990s, D. F. Specht first proposed the concept of generalized regression neural networks. GRNN is another variant of radial basis function. GRNN is extremely robust and fault-tolerant, and it has a flexible network structure and strong nonlinear reflection capabilities, so it can handle nonlinear problems well. Compared with RBF, its learning speed and approximation ability are relatively stronger, and the regression surface of GRNN convergence has a larger sample size. The generalized regression network structure diagram is shown in Fig. 4.13.

As shown in Fig. 4.13, the components of the generalized regression neural network mainly include summation layer, pattern layer, output layer, and input layer, as described below:

(1) Input layer

The number of neurons in the input layer is equal to the dimension of the input vector, and the input layer directly transfers the variables to the pattern layer.

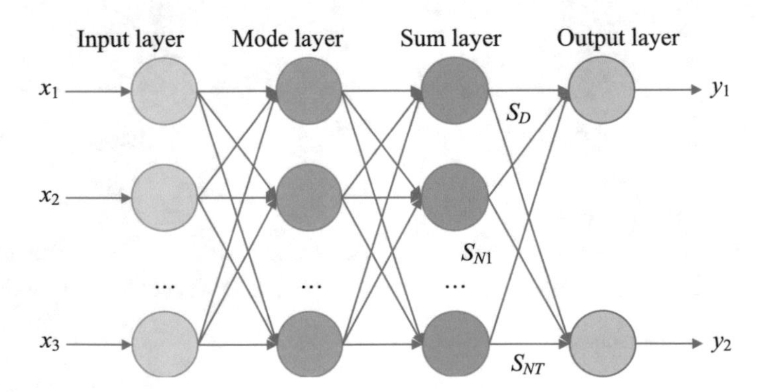

Fig. 4.13 Generalized regression network structure diagram

(2) Pattern layer

The number of neurons in the pattern layer is equal to the number of learning samples n, and the transfer function of the neurons in the pattern layer is:

$$p_i = \exp\left[-\frac{(X - X_i)^{\mathrm{T}}(X - X_i)}{2\sigma^2}\right] \tag{4.13}$$

where X is the network input variable; X_i is the sample corresponding to the i-th neuron, the value range of neuron i is $1 - n$, and σ is the function width coefficient.

(3) Summation layer

Two types of neurons are used for summation in the summation layer. The first type of calculation formula is:

$$\sum\nolimits_{i=1}^{n} \exp\left[-\frac{(X - X_i)^{\mathrm{T}}(X - X_i)}{2\sigma^2}\right] \tag{4.14}$$

It performs arithmetic summation on the output of all pattern layer neurons, and the transfer function is:

$$S_D = \sum_{i=1}^{n} p_i \tag{4.15}$$

The second type of calculation formula is:

$$\sum_{i=1}^{n} Y_i \exp\left[-\frac{(X - X_i)^{\mathrm{T}}(X - X_i)}{2\sigma^2}\right] \tag{4.16}$$

which weighted sums the outputs of all pattern layer neurons, the connection weight between the ith neuron in the pattern layer and the jth molecular summation neuron in the summation layer is the jth element in the ith output sample Y_i, and the transfer function is:

$$S_{Nj} = \sum_{i=1}^{n} y_{ij} p_{i,}, j = 1, 2, \ldots, k \tag{4.17}$$

(4) Output layer

The number of neurons in the output layer is equal to the number of output vectors k in the learning sample. Each neuron divides the output of the summation layer. The result is:

$$y_j = \frac{S_{Nj}}{S_D}, j = 1, 2, \ldots, k \tag{4.18}$$

To determine the error, the actual output on the output layer is compared with the desired output. Based on this error value, the weight matrix between the input and output layers will be updated. Continue to use SUMFC of blurred images as input to the network.

(2) Generalized regression neural network theory

Unlike traditional networks, the learning algorithm of the generalized regression neural network does not adjust the connection weights between neurons during training, but changes the smoothing parameters, thereby adjusting the transfer function of each unit in the model layer to obtain the best regression estimation result. The generalized regression neural network's $f(x, y)$ theoretical basis is nonlinear regression analysis. The regression analysis of the independent variable Y relative to the independent variable x is actually to calculate y with the largest probability value. Suppose the joint probability density function of a random variable x and random variable y is $f(x, y)$. If the observed value of x is X, then the regression of y relative to X is the conditional mean:

$$\hat{Y} = E(y/X) = \frac{\int_{-\infty}^{\infty} yf(X, y)\mathrm{d}y}{\int_{-\infty}^{\infty} f(X, y)\mathrm{d}y} \tag{4.19}$$

$\hat{Y}$ is the predicted output of Y under the condition that the input is X. Using the Parezn nonparametric estimation [92], the density function can be estimated from the sample data set $\{x_i, y_i\}_{i=1}^{n}$:

$$\hat{f}(X, y) = \frac{1}{n(2\pi)^{\frac{p+1}{2}}\sigma^{p+1}} \sum_{i=1}^{n} \exp\left[-\frac{(X - X_i)^{\mathrm{T}}(X - X_i)}{2\sigma^2}\right] \exp\left[-\frac{(X - Y_i)^2}{2\sigma^2}\right] \tag{4.20}$$

In Eq. (4.20), X_i and Y_i are the sample observation values of the random variables x and y; n is the sample size; p is the dimension of the random variable x; σ is the width coefficient of the Gaussian function, which is called smoothness factor here.

Use $\hat{f}(X, y)$ instead of $f(X, y)$ in Eq. (4.19) and exchange the order of integration and addition to get:

$$\hat{Y}(X) = \frac{\sum_{i=1}^{n} \exp\left[-\frac{(X-X_i)^{\mathrm{T}}(X-X_i)}{2\sigma^2}\right] \int_{-\infty}^{\infty} y \exp\left[-\frac{(Y-Y_i)^2}{2\sigma^2}\right]\mathrm{d}y}{\sum_{i=1}^{n} \exp\left[-\frac{(X-X_i)^{\mathrm{T}}(X-X_i)}{2\sigma^2}\right] \int_{-\infty}^{\infty} \exp\left[-\frac{(Y-Y_i)^2}{2\sigma^2}\right]\mathrm{d}y} \tag{4.21}$$

Because of $\int_{-\infty}^{\infty} z\mathrm{e}^{-z^2}\mathrm{d}z = 0$, calculate the two integrals and get the output of the network $\hat{Y}(X)$ as

$$\hat{Y}(X) = \frac{\sum_{i=1}^{n} Y_i \exp\left[-\frac{(X-X_i)^{\mathrm{T}}(X-X_i)}{2\sigma^2}\right]}{\sum_{i=1}^{n} \exp\left[-\frac{(X-X_i)^{\mathrm{T}}(X-X_i)}{2\sigma^2}\right]} \tag{4.22}$$

The estimated value $\hat{Y}(X)$ is the weighted average of all sample observations Y_i, and the weighting factor of each observation Y_i is the index of the square of the Euclidean distance between the corresponding sample X_i and X. When the smoothing factor σ is very large, $\hat{Y}(X)$ approximates the mean of all sample dependent variables. In contrast, when the smoothing factor σ tends to 0, $\hat{Y}(X)$ is very close to the training sample. When the point to be predicted is included in the training sample set, the predicted value of the dependent variable obtained by the formula will be very close to the corresponding dependent variable in the sample. Once the points that are not included in the sample are encountered, it is possible that the prediction effect will be very poor. This phenomenon indicates that the generalization ability of the network is poor. When the value of σ is moderate and the prediction value of $\hat{Y}(X)$ is obtained, the dependent variables of all training samples are taken into account, and the dependent variables corresponding to the sample points with the closest prediction points are given greater weight.

(3) Generalized regression neural network training

First, we need to train the neural network. The training selects eight tag images, and the selection range of the fuzzy length is from 2 to 15, and the step size is 1. Calculate the SUMFC of each image in a total of 112 training samples. Control the resulting error is less than 0.01. After training, use the trained neural network to estimate the fuzzy length.

The collection of tag image information is done by CCD. First, the Gabor filter is used to obtain the blurred angle of the image, and the image is rotated by the obtained angle information to change the blur direction of the image to the horizontal direction. Then, the Fourier coefficients of the horizontally blurred image in the spectrum domain are used as input to train the neural network to estimate its fuzzy length, as shown in Fig. 4.14 for the convergence curve of the neural network.

These data will be applied to the establishment and training of the neural network to obtain the corresponding model, through which the fuzzy length can be predicted. By comparing and analyzing the preset data and the predicted data, you can determine the performance of the model and determine whether it can obtain accurate prediction results.

4.2.4 De-Motion Blur Analysis

In this study, after obtaining the estimated blur kernel through the estimated blur parameters, the Wiener filter is used to restore the blurred image. In order to verify the effectiveness of the method in this paper, the experiment uses motion blur with

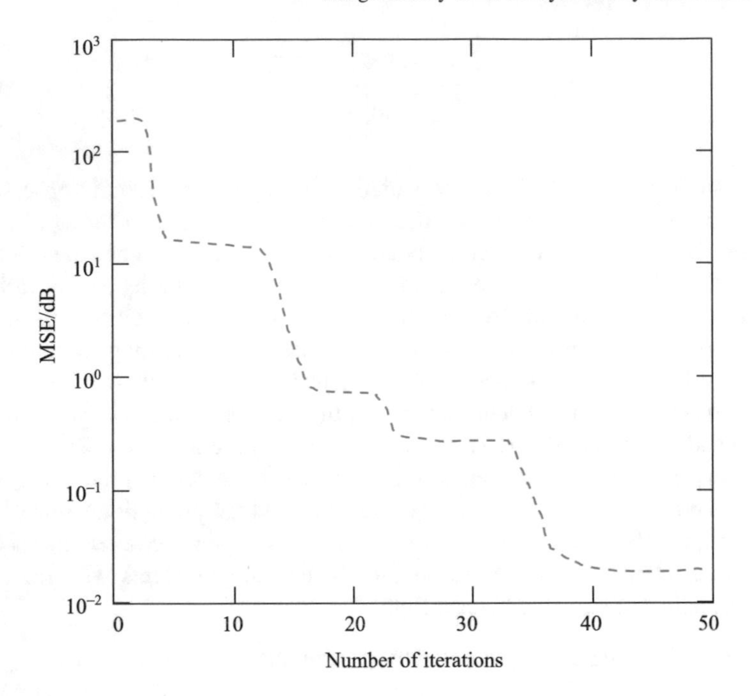

Fig. 4.14 Neural network convergence curve

different blurred angles and blurred lengths on multiple images and compares them with the methods described in this paper by traditional methods. Taking the mean square error (MSE) of the image as the measurement standard, the root mean square error is defined as follows:

$$\mathrm{MSE} = \frac{1}{n \times m} \sum_{i=1}^{n \times m} (f(x, y) - g(x, y))^2 \tag{4.23}$$

In the above formula, $f(x, y)$ and $g(x, y)$ represent the pixel values at the (x, y) position of the original and restored images of size $m \times n$. According to the meaning of MSE in the above formula, when MSE $= 0$, it means that the two test images are completely consistent, and the restoration effect at this time is the best; when MSE > 0, it means that the two images are not consistent, and the difference between the images increases as the MSE value increases. When MSE is used to evaluate the quality of image restoration, the smaller the value, the better the restoration effect.

As shown in Fig. 4.15, the acquisition method of (a), (b), (c), and (d) is to rotate the turntable first and the side with the RFID tag, and the matching pattern facing the horizontal CCD camera is taken in a stationary state. The experiment uses part of the tag image obtained through the experiment as a reference and simulates the actual situation to make it have different blurred lengths and blurred angles. The applied blur and deblurring conditions are shown in Fig. 4.16. Figure (a) and (e) are the experimental diagram and restoration diagram with a blurred length of 10 and a

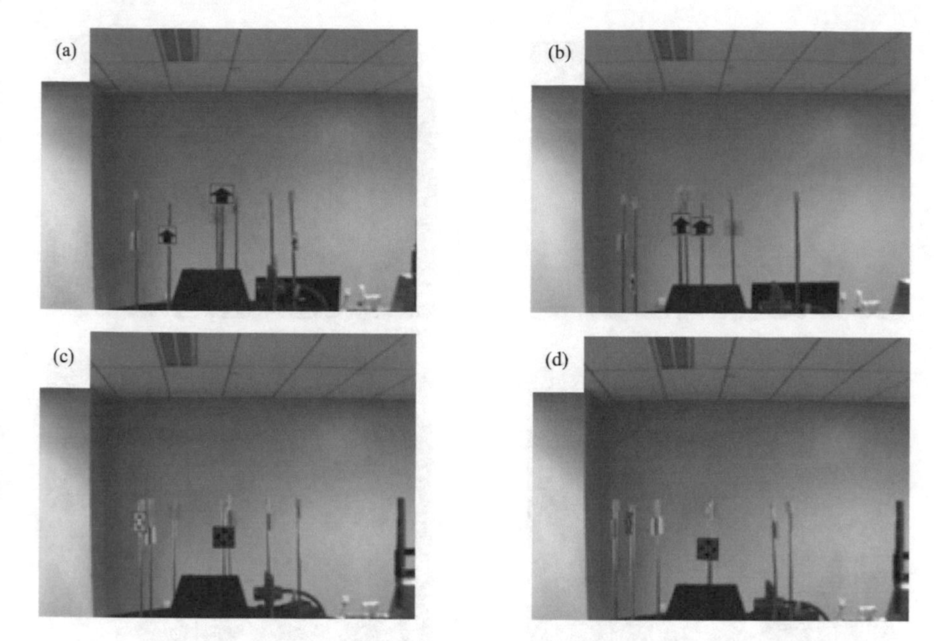

Fig. 4.15 Original image of the tag used in the experiment

blurred angle of 15°, respectively; Figure (b) and (f) are the experimental diagram and restoration diagram with the blurred length of 15 and the blurred angle of 13°, respectively; Figure (c) and (g) are the experimental diagram and restoration diagram with a blurred length of 30 and a blurred angle of 40°, respectively; Figures (d) and (h) are an experimental diagram and a restoration diagram with a blurred length of 30 and a blurred angle of 45°, respectively.

Images with different blurred lengths and angles are used as input to the proposed estimation scheme. The Gabor filter is used to calculate the blurred angle of each image, and the neural network is used to calculate the length of each image. According to the data analysis obtained from the experiment, it can be found from Table 4.5 that the Gabor filter estimates the angles at various angles ideally. After estimating the blurred angle, the neural network is used to estimate the blurred length. It can be found that the estimated blurred length is also very close to the original value.

A PSF model is established based on the predicted (θ, L). In this study, a Wiener filter was selected to restore the blurred image, and the restoration results were evaluated using MSE as the evaluation index. The results are shown in Table 4.5. From Table 4.5, we can see that the method of this paper is used to restore the motion blurred image and the MSE between the restored image and the original image is small, which shows that the method of this paper can effectively remove the motion blur in the image.

This chapter proposes a method to estimate the blur kernel to achieve the motion blur when the three-dimensional coordinate measurement of the RFID tag cannot obtain a clear image of the tag due to motion blur. This method is based on the prior

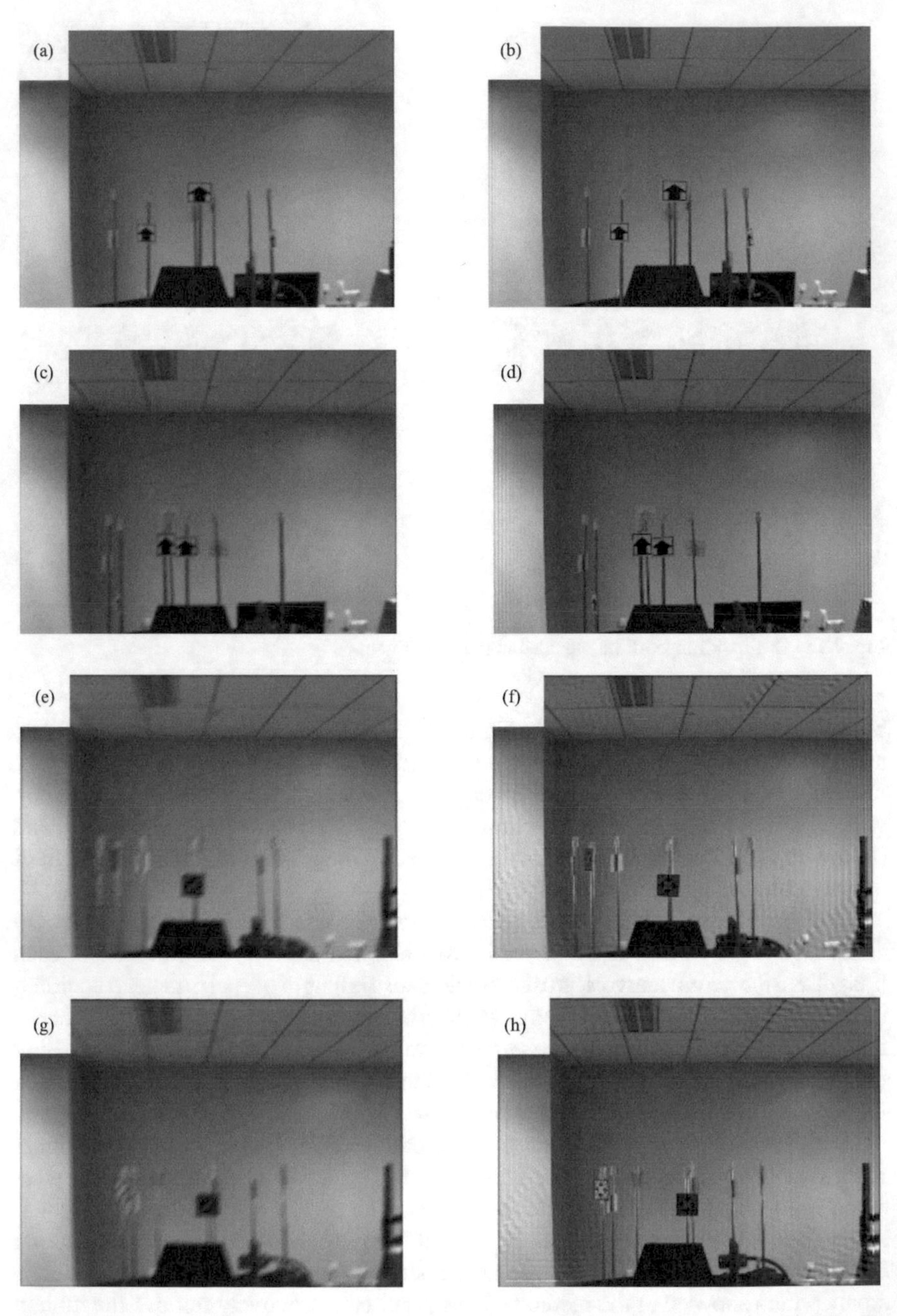

Fig. 4.16 Tag blur images (**a**), (**c**), (**e**), and (**g**) and corresponding recovery images (**b**), (**d**), (**f**), and (**h**)

Table 4.5 Blurred parameter estimation and MSE calculation result

Image	Blurred length L	Blurred angle θ	Estimated length L	Estimated angle θ	MSE
(a)	10	15	10	14	1.22E-02
(b)	15	30	14	29	3.56E-02
(c)	30	40	29	39	2.29E-02
(d)	30	45	29	47	2.32E-02

blur estimation of the motion-blurred image and restores the image based on the obtained estimated PSF. According to the characteristics of motion blurred image in frequency domain, this chapter uses Gabor filter to estimate the blurred angle. The generalized regression neural network is used to estimate the blurred length according to the nonlinear relationship between the blurred length of the motion blurred image and the Fourier coefficient of its spectrogram. After obtaining the blur parameters, the Wiener filtering method is used to restore the blurry image. The research results show that the method proposed in this chapter can effectively repair the blur image formed by the relative motion between the tag and the camera during the three-dimensional coordinate measurement of the RFID tag. The algorithm proposed in this study improves the accuracy of image-based RFID three-dimensional coordinate measurement and lays a good research foundation for subsequent acquisition of three-dimensional coordinate distribution of tags.

4.3 RFID Tag Positioning Method

4.3.1 Image Matching Overview

In the process of computer vision recognition, it is often necessary to compare two or more images of the same scene obtained by different sensors or the same sensor at different times and under different imaging conditions to find a common scene in the group of images. It is based on the known pattern to find the corresponding pattern in another picture, which is called image matching [8, 9]. The so-called digital image matching is simply to find the best transformation of the corresponding point of one image to another image for the digital image.

The similarity evaluation ability of images has always been the basic core of image processing and computer vision tasks (such as target recognition, classification, and texture classification). Image matching is a task of the machine. Its goal is not to obtain a satisfactory quality of experience, but to extract unique features from the image location. These features can be matched when the same scene is captured under different transformations. Compared to perceptual attributes, these features are designated as invariance to changes in geometry and luminosity. Therefore, in the image matching technology, there is also the problem of optimization of perceptual excitation.

Image matching techniques can be divided into two categories. The first aspect is the image representation method, and the second aspect is to define an appropriate similarity metric to compare the images in the selected representation space. Different representations and similarity measures are introduced in image matching technology, and in particular, image representations are also selected for each application. However, finding the appropriate similarity measure will make the problem more complicated.

As shown in Fig. 4.17, image matching involves several steps, specifically input image, preprocessing, information extraction, matching image, and result output. Because the method selected in the image matching process is different, the steps of each algorithm are not completely consistent, but there is no obvious difference in the general process.

Because the characteristics of each image are different, for example, the focus, line, bar, skeleton, edge, etc. of each image are not completely consistent, so it can be greatly reduced even if the algorithm's adaptability is improved. The algorithm is less sensitive to small disturbances, but more dependent on the extracted features. The key

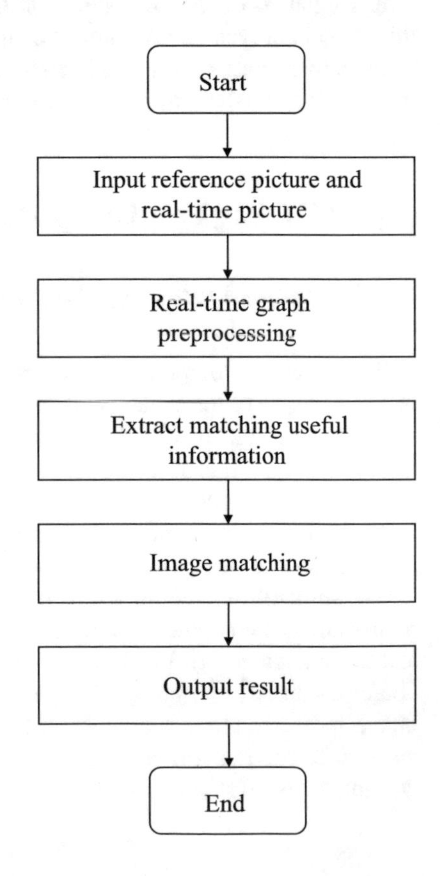

Fig. 4.17 Image matching process

of the algorithm is to determine the features that are easy to distinguish and identify and determine the matching position through the similarity of each feature set.

4.3.2 Image Threshold Segmentation Technology

Determining the characteristics of the tag and separating the tag from the background are important contents of tag positioning. If there is a significant difference between the tag background and the grayscale set of the tag, and the two grayscale sets can be segmented by the threshold, then the image can be segmented by the threshold to find the tag in the image.

For example, if a method of threshold image segmentation is selected, a corresponding threshold value needs to be preset in conjunction with the gray value of the tag, and the number of threshold values may be one or more. Usually combined with the display requirements, the target and background are represented by 0 and 255, respectively. Because the background and the target may not only exist in the two gray sets, it is necessary to set several thresholds during the process of extracting the target. If the value of the pixel is within the threshold, then the system will be recognized as the target, and on the contrary, the system recognized as a background can be described by the following formula:

$$g(x, y) = \begin{cases} Z_E, & T_1 < f(x, y) < T_2 \\ Z_B, & \text{Othercases} \end{cases} \tag{4.24}$$

Among them, Z_E and Z_B represent the grayscale of the target and background in the image, respectively, and $[T_1, T_2]$ is the selected threshold range.

For threshold segmentation, if the extraction target or background has only small grayscale fluctuations, then better extraction effects can be exerted. Only the threshold value needs to be understood as a fixed value. In such a case, only setting a reasonable threshold can achieve the purpose of image segmentation. However, the target and background usually have large grayscale fluctuations, especially if the environment is more complex, then the grayscale of the target and part of the background may be very close, so part of the background content may be judged as the target, and the problem of misidentification may occur, which affects subsequent processing. In such cases, it is necessary to set the corresponding threshold value in accordance with the difference in the image position. There are several common methods to choose the best threshold:

(1) Maximum variance threshold

The basic idea of the maximum variance threshold is to set a certain threshold and divide the histogram to obtain two groups. The method for determining the threshold is to make the two groups have the largest variance. A histogram is a method of expressing an image in a statistical manner. A series of vertical stripes with different heights represent the data distribution. The grayscale histogram belongs to a grayscale function. The number of pixels in the grayscale set describes the frequency of a certain grayscale (that is, the number of 0–255 pixels in the image).

The grayscale histogram describes the proportion of various grayscales in the image (for an image with a pixel depth of 8 bits, a total of 256 values ranging from 0 to 255) in the entire image.

(2) Double-peak method selection threshold

The principle of this method is that the background and the foreground together constitute an image, or the two groups of colors constitute an image, and the gray levels of the pixels of the two groups appear as peaks, and the valley corresponds to the threshold. After the threshold is determined by this method, the image can be segmented. This method is relatively simple, and the threshold can be roughly determined through observation, but this method cannot reflect the image details well, so there are certain limitations.

4.3.3 Edge Detection Algorithm

The image contains many features, and the edge is the most basic one among the features. The image can be effectively identified through the edge. The edge belongs to a key parameter that interprets the image, identifies the target, or describes the image. It is located between the area, the target, and the background. Edges are the basis of image and texture features. The edge detection method can be understood as a calculation for points, by which image processing can be understood as signal processing. There is a lot of information in the image, and the edge belongs to a tight description of the image. The edge information is extremely critical to the image, so in the field of preprocessing, edge detection is of great significance.

(1) Edge

The image contains many features, and the edge is just one of them. The edges can outline the image, allowing people to understand the image in a direct way. When there are significant changes in the image, the pixel set of these changes is the edge, which is related to the object, but it is not consistent. If the color of the background and the boundary of the object are the same, it is difficult to identify the edge. The discontinuity of grayscale can be described by the edge. In essence, the grayscale abrupt boundary is the edge, which can explain the formation of one area and the end of another area.

In the fields of machine vision, pattern recognition, image segmentation, and boundary detection, edges have important significance, and they belong to the basis of contour detection and boundary detection. There are a lot of edges between primitives, objects, backgrounds, and objects, so edges are the key features when segmenting images. The image features of the edges appear as partially discontinuous grayscale, that is, the most obvious changes in brightness. Generally speaking, the edge has a gentle grayscale change, and the grayscale on both sides has a more obvious change.

(2) Gradient

The step caused by the change in gray level can be described by the gradient in the mathematical field.

The edges of objects are caused by discontinuous changes in grayscale. The classic edge extraction method is called the local operator method. This method selects a certain range and examines the gray levels of all pixels in the range to determine the gray change rule. This method is relatively simple.

If there is a certain pixel in the image boundary, then the field of this pixel is also called a certain gray level change band. The magnitude and direction of the gray change rate on the gradient vector are the most effective features for this change.

The edge detection operator needs to check all pixel fields, describe the gray change rate in a quantitative way, and determine the direction of change. The method used mainly needs to be convolution based on the direction derivative by modulus.

In the process of processing images, if you choose the integration method, you can't get a clear edge, but it has the opposite effect on differentiation. Therefore, differentiation is often used in the process of edge detection. Among the various differential processing methods, the most widely used is the gradient method. The gradient of the quantity field means.

Quantity field $u = u(x, y, z)$ and the vector whose magnitude is the maximum value of the directional derivative at a certain point and whose direction is the maximum value of the directional derivative is called the gradient of the quantity field.

$$\vec{G}(u) = \frac{\partial u}{\partial x}\vec{i} + \frac{\partial u}{\partial y}\vec{j} + \frac{\partial u}{\partial z}\vec{k} \tag{4.25}$$

Let the image be $f(x, y)$, the gradient vector at points x, y is:

$$\vec{G}[f(x, y)] = \begin{bmatrix} \frac{\partial f}{\partial x} \\ \frac{\partial f}{\partial y} \end{bmatrix} \tag{4.26}$$

It can be found by definition that there are two main characteristics of the gradient:

(1) The vector $\vec{G}[f(x, y)]$ is the direction pointing to the maximum increase rate of $f(x, y)$;
(2) If $\vec{G}[f(x, y)]$ is used to express the magnitude of $grad[f(x, y)]$, then

$$G[f(x, y)] = \max\left\{\vec{G}[f(x, y)]\right\} = \sqrt{\left(\frac{\partial f}{\partial x}\right)^2 + \left(\frac{\partial f}{\partial x}\right)^2} \tag{4.27}$$

It can be found from the analysis of the above formula that it belongs to a scalar function, and the value is always positive. Since this formula will appear frequently in the following text, it will be referred to as the gradient module in the subsequent discussion.

(3) Edge detection based on the Canny algorithm

Canny edge detection was proposed by John Canny in 1986. This detection method can extract valuable structural information from the image. Since the extracted information is less, it can greatly reduce the processing amount. At present, there are many visual systems. The method is applied. Canny found that the requirements for edge detection are similar on different vision systems. Therefore, an edge detection technology with broad application significance can be realized. The general standards for edge detection methods are:

(1) Maintain a low error rate in the process of detecting edges, which means that you need to find the edges in the image as accurately as possible.
(2) The center of the edge needs to be accurately determined by detection.
(3) The number of times of marking an edge can only be 1 to reduce the interference of noise on detection as much as possible and to avoid false edges. To meet these requirements, Canny used a variational method. The optimal function in the Canny detector is described by the sum of four exponential terms, which can be approximated by the first derivative of the Gaussian function. Among the edge detection methods, the most rigorously defined method is the Canny method, which can detect the edges in the image stably and reliably. Because this method is not complicated and meets three standards, it has become the edge in recent years and one of the preferred methods in the detection process.

The gradient of the Canny edge detection operator is calculated using the derivative of the Gaussian filter, and the edge appears at the local maximum of the gradient. The steps of the Canny algorithm are as follows:

(1) Use Gaussian function to smooth the image and filter out noise. The formula of the Gaussian function is:

$$G(x, y) = \frac{1}{2\pi\sigma^2} \exp\left[-\frac{1}{2}\left(\frac{x^2 + y^2}{2\sigma^2}\right)\right] \tag{4.28}$$

The noise is removed by the convolution of the Gaussian function and the image $f(x, y)$, and the formula to calculate the two-dimensional convolution is as follows:

$$f(x, y)' = \nabla G(x, y) * f(x, y) \tag{4.29}$$

Calculate the filtered edge strength and direction, and use the threshold to detect the edge. Decompose the two-dimensional convolution template of $\nabla G(x, y)$ into two one-dimensional filters:

$$\frac{\partial G(x, y)}{\partial x} = kx \cdot \exp\left(-\frac{x^2}{2\sigma^2}\right)\exp\left(-\frac{y^2}{2\sigma^2}\right) \tag{4.30}$$

$$\frac{\partial G(x, y)}{\partial y} = ky \cdot \exp\left(-\frac{x^2}{2\sigma^2}\right)\exp\left(-\frac{y^2}{2\sigma^2}\right) \tag{4.31}$$

After convolution of these two templates with $f(x, y)'$ we can get

$$E_x = \frac{\partial G(x, y)}{\partial x} * f(x, y)' \tag{4.32}$$

$$E_y = \frac{\partial G(x, y)}{\partial y} * f(x, y)' \tag{4.33}$$

Then the expression of the obtained gradient and direction is:

$$A(x, y) = \sqrt{E_x^2 + E_y^2} \tag{4.34}$$

$$a(x, y) = \arctan\left[\frac{E_x(x, y)}{E_y(x, y)}\right] \tag{4.35}$$

(2) Determine whether the pixel is an edge point. In the process of judging whether a pixel is an edge, there are three main reference conditions. First, the edge intensity of (x, y) is greater than the edge intensity of two adjacent pixels along the gradient direction; second, the direction difference between the two adjacent pixels in the gradient direction of the pixel is less than 45°; third, the maximum value of the edge intensity in the 3×3 neighborhood centered on the pixel is smaller than the threshold that has been set.

(4) Edge detection based on LOG algorithm

The LOG (Laplacian of Gaussian) operator is based on the Laplacian and Gaussian functions [16]. Scholars such as Torre found that the smooth function of Gaussian function is close to optimal. Scholars such as Marr chose Gaussian in the process of smoothing the image and then chose the Laplace operator to detect the edge based on the second derivative zero-crossing point. This method is also called the LOG operator. The Laplacian operator is a second-order differential operator, which has no dependence on the edge direction. It is scalar and has the property of rotation invariance. It is often used to extract the edge of the image in image processing. The expression is:

$$\nabla^2 f = \frac{\nabla^2 f}{\nabla x^2} + \frac{\nabla^2 f}{\nabla x^2} \tag{4.36}$$

The approximate formula for digital images is:

$$\nabla^2 f(x, y) = f(x + 1, y) + f(x - 1, y) + f(x, y + 1) + f(x, y - 1) - 4f(x, y) \tag{4.37}$$

Since the Laplace operator is a second-order differential operator, it is very sensitive to noise in the image. The LOG operator belongs to a detection operator improved based on the classic operator. The signal-to-noise ratio needs to be determined during the calculation of the optimal filter.

The LOG operator has three main advantages. First, the calculation speed is relatively fast; second, the Gaussian filter can achieve the best in the frequency domain and the space domain; third, it has the excellent anti-jamming ability, good continuity, strong accuracy, and can reduce the contrast. The boundary is accurately presented. However, this method also has certain limitations. For example, if the width of the operator is greater than the width of the boundary, the zero-crossing slope will merge, which will cause the problem of losing some boundary details.

$$LOG = \nabla^2 G(x, y) = \frac{1}{\pi\sigma^4}\left(\frac{x^2 + y^2}{2\sigma^2} - 1\right)\exp\left(-\frac{x^2 + y^2}{2\sigma^2}\right) \tag{4.38}$$

In the process of detecting step edges by zero-crossing, the LOG operator is the best operator at this stage, but in fact, the zero-crossing points are not all edge points. So when determining the zero-crossing point, it is necessary to verify whether it is accurate. In addition, in essence, the facet model is also a method for edge detection based on second-order differential zero-crossings.

4.4 3D Space RFID Tag Location

The experimental environment of the three-dimensional RFID tag positioning in this study is shown in Fig. 4.18. Figure 4.18a is the matching pattern attached to the bottom of the tag holder. The whole positioning process is based on the three-dimensional multi-tag positioning system described in Chap. 2.

First, the system will grayscale the acquired pictures, and the deblur process is performed on the acquired image by the method of de-motion blur described in Chap. 3. As shown in Fig. 4.18b, it is the image for matching positioning captured by the vertical CCD camera after deblur process. The threshold segmentation of the picture highlights the need to obtain the tag holder pattern and the center pattern of the turntable, as shown in Fig. 4.19.

Then, the system determines the center position of the turntable based on the image matching technology of the center of the turntable, as shown in Fig. 4.19b. At the same time, the image matching of the tag holder pattern is also performed. The ratio of each pixel in the image to the actual length can be determined from the vertical distance between the turntable and the vertical CCD and the CCD camera itself and the field of view size. The horizontal camera and vertical camera have been calibrated

(a) Matching pattern of tag holder base (b) Images used for matching in the experiment

Fig. 4.18 Pattern matching of the tag holder base

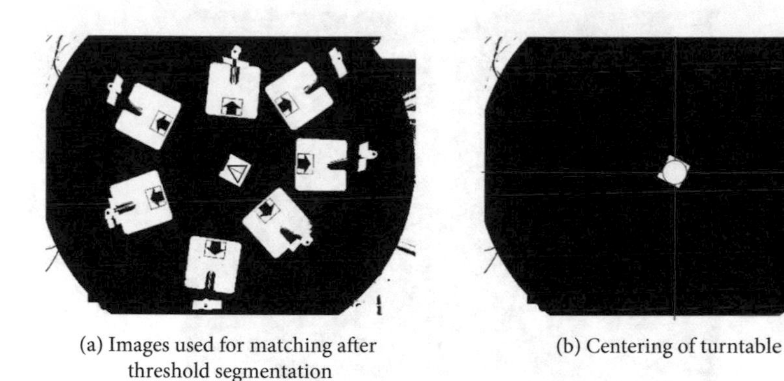

(a) Images used for matching after threshold segmentation (b) Centering of turntable

Fig. 4.19 Tag positioning image after threshold segmentation

before the experiment, and their scale factors have been obtained before the experiment. The resolution of both horizontal and vertical cameras is 1280 × 964 pixels, the radius of the turntable is 0.65 m, each pixel in the field of view of the vertical camera corresponds to the surface length of the turntable is 1 mm, and each pixel in the field of view of the horizontal camera corresponds to the surface length of the turntable. It is 1 mm, which can determine the distance between each tag and the center of the turntable, as shown in Fig. 4.20a. Figure 4.20b shows the distance from each tag holder to the center of the turntable.

The principle of horizontal coordinate measurement is shown in Fig. 4.21. The method for determining the position of the tag is shown in Fig. 4.21a. Use the center of the turntable as the origin of the polar coordinate system, and use one of the drawings as the template. The template to the last tag bracket pattern is given numbers in sequence. Rotate the turntable, when the *n*th tag holder pattern rotates to the alignment, obtain its rotation angle θ_n, and the distance r_n of the pattern has also been obtained. Repeat the above rotation steps until all the pattern information has been obtained. So far, in the polar coordinate system with the center of the turntable

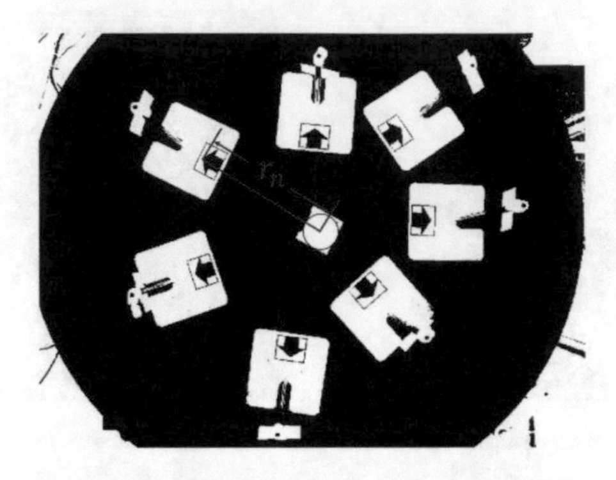

(a) Distance from the tag holder pattern to the center of the turntable

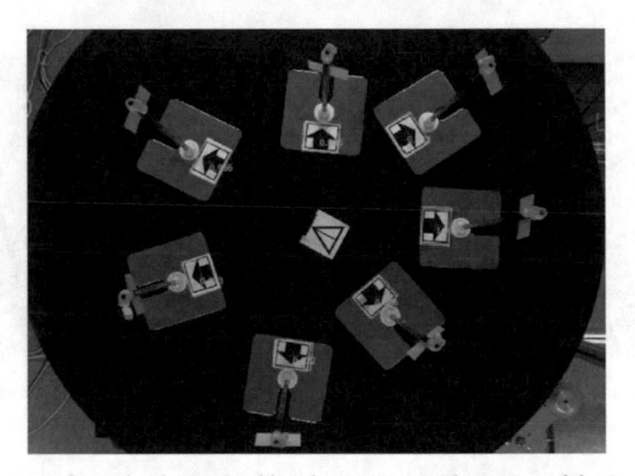

(b) Distance from the determined bracket pattern to the center of the turntable

Fig. 4.20 The distance between the tag holder pattern and the center of the turntable

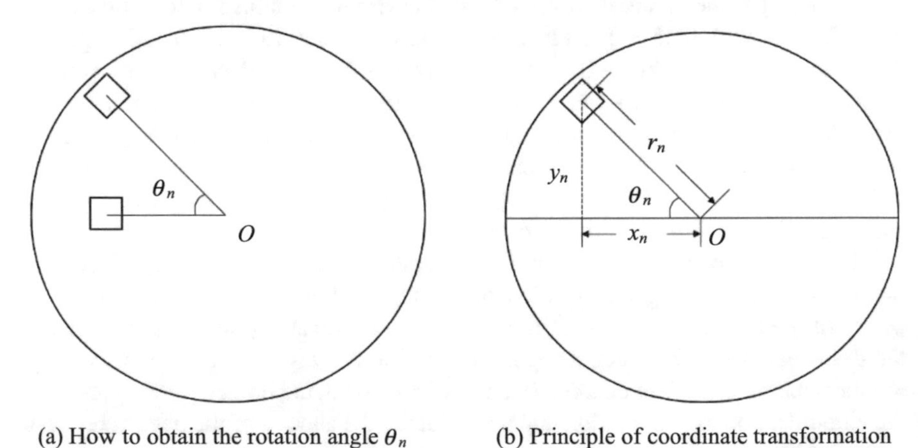

(a) How to obtain the rotation angle θ_n (b) Principle of coordinate transformation

Fig. 4.21 Horizontal coordinate measurement principle

as the origin, the distance r from the origin and the rotation angle θ_n of any pattern n have been obtained.

As shown in Fig. 4.21b, the polar coordinates that have been acquired can be converted into two-dimensional horizontal coordinates. For the nth tag bracket pattern, its two-dimensional coordinates $(r_n \cos\theta_n, r_n \sin\theta_n)$ can be obtained by its r_n and θ_n.

After the horizontal position of all tags is measured, the turntable will rotate the corresponding angle according to the horizontal position of each tag, so that the tag can face the horizontal camera. The horizontal camera will automatically focus according to the position of the tag to be measured to find the best collection point to obtain the height information of the tag.

For the nth tag, when the front face rotates to face the horizontal camera, the camera captures the image. Like the process of obtaining two-dimensional horizontal coordinates, the collected image is processed by grayscale and deblurring. The system has entered the image information of the tag and performs image matching on the processed image through the edge detection algorithm. The tag information is matched in the image to determine the area where the tag is located, as shown in Fig. 4.22.

The ratio of each pixel to the actual length in the processed image can be determined by the distance of the tag relative to the horizontal camera. Since the conveyor belt is fixed with respect to the horizontal camera, the distance between the tag and the horizontal camera can be obtained by subtracting the distance r_n of the corresponding turntable center from the center of the turntable. The center of the field of view of the horizontal camera has been determined, and the height information can be obtained by calculating the pixel distance between the center of the field of view and the identified tag area. As shown in Table 4.6, the final tag three-dimensional coordinates $(r_n \cos\theta_n, r_n \sin\theta_n, I_n)$ have been obtained, and the tag three-dimensional coordinate positioning is completed [96].

One set of three-dimensional coordinates is randomly selected for drawing, and the spatial distribution is shown in Fig. 4.23.

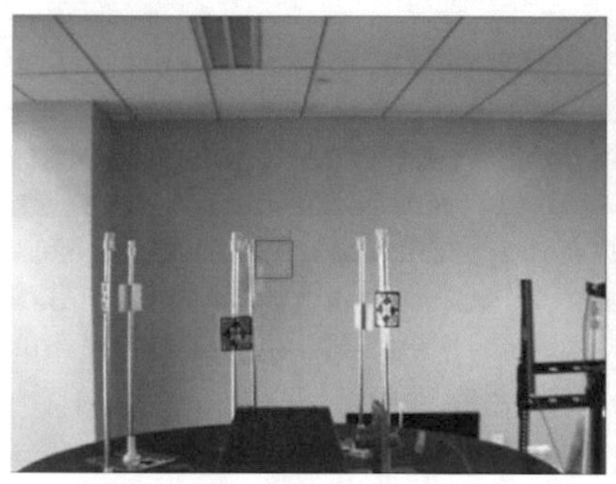
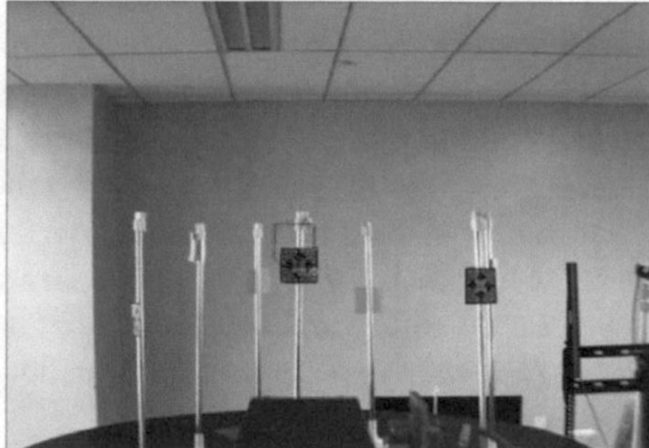

Fig. 4.22 Edge detection determines the area where the tag is located

Table 4.6 Multi-tag three-dimensional spatial coordinate distribution

x_1/m	y_1/m	z_1/m	…	x_7/m	y_7/m	z_7/m
0.813	0.262	0.216	…	0.942	0.535	0.265
0.963	0.568	0.111	…	0.431	0.752	0.158
0.188	0.967	0.139	…	0.735	0.596	0.206
⋮	⋮	⋮		⋮	⋮	⋮
0.937	0.922	0.203	…	0.116	0.342	0.035
0.606	0.145	0.289	…	0.485	0.859	0.192
0.027	0.957	0.077	…	0.900	0.657	0.173

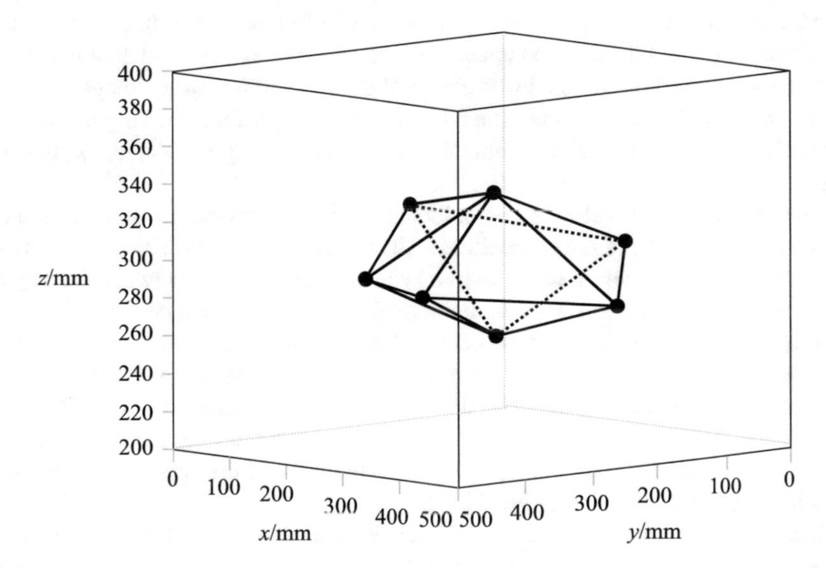

Fig. 4.23 Three-dimensional spatial distribution of RFID tags

This system obtains the horizontal multi-tag image of the tag through the vertical camera and the vertical multi-tag image of the multi-tag through the horizontal camera, and the position matching of multiple tags is performed through template matching, so as to effectively obtain the three-dimensional coordinates of the tags. Through a lot of experiments, it is shown that the experimental results of this system are basically consistent with the actual measured values, as shown in Table 4.7, which proves the reliability and applicability of the method proposed in this chapter from the perspective of experimental verification.

Table 4.7 Experimental measurement results are not compared with actual measurement values

	x_1/m	y_1/m	z_1/m	…	x_7/m	y_7/m	z_7/m
Experimental measurements	0.813	0.262	0.216	…	0.942	0.535	0.265
Actual measured value	0.817	0.263	0.219	…	0.947	0.538	0.268
Error/%	0.49	0.38	1.4	…	0.53	0.56	1.2

4.5 Novel Reverse Design Method of Tag Antenna Based on Image Analysis

4.5.1 Introduction

RFID (radio frequency identification) technology is a non-contact identification technology, which has the advantages of large information capacity, strong adaptability, and high precision. RFID technology is widely used in daily life and gradually penetrated all fields of people's study, life, and work. In order to meet the needs of people's ever-changing and rapid growth of information, RFID technology has been recognized by more and more people and has become a hot research field.

UHF RFID technology is widely used due to its long reading range and low cost. The RFID system includes an antenna, a reader, a tag, and a management system. The most important part is the RFID tag. The tag antenna receives the RF signal from the reader and transmits the chip data to the reader through backscattering. So, the tag performance depends mainly on the antenna. In UHF RFID tags, not only conjugate matching of chip impedance and antenna impedance is required, but also the antenna has good directivity.

The design of the tag antenna is based on the application requirements of the antenna, and the different requirements are derived from the actual application environment of the tag. At present, the main research contents of UHF RFID tags are focused on RFID system application design [6], tag chip research, and antenna optimization design. In terms of antenna design, it mainly focuses on the improvement of antenna performance [3, 5], such as antenna and chip impedance matching [4], increasing the tag bandwidth, and reducing the return loss. Although there are many UHF antenna structures, the main research object is the dipole tag antenna.

UHF antennas are basically designed to miniaturize or resist metal interference on the basis of dipole antennas or to improve antenna performance [5, 6]. Yang proposes a polarization diversity antenna composed of two dual-planar inverted-F antennas [7]. The antenna can work with a linearly polarized antenna with any polarization direction and the antenna can operate normally on a metallic medium. However, the return loss of the antenna is very low, which requires high antenna incident power. Hamani designed a new UHF antenna that covers the entire UHF band and achieves good read and write distances when working on metal plates [8]. The current research mainly focuses on the performance analysis of the antenna in practical applications, and the research on the antenna design method is less. Multiple performances should be considered when designing the antenna, but the antenna performances cannot be fully improved.

The basic model used in this paper is a half-wave dipole antenna, and an impedance matching loop is added to the antenna. This model has a simple structure and strong applicability. Antenna performance can be varied by adjusting the basic antenna geometry. The antenna can quickly adjust the impedance value of the tag antenna by changing the parameters of the center impedance matching loop to achieve the purpose of matching with the tag chip. Through the changes and analysis of the antenna parameters, the key factors affecting the antenna performance indicators can be summarized. Referring to the existing analysis results, the tag antennas that meet the requirements of different applications can be obtained through corresponding parameter optimization.

In this chapter, the dipole antenna model was analyzed theoretically, and the important parameters that affect the antenna performance were found out. Taking the bent dipole antenna as an example, this paper analyzed the parameters of the dipole tag antenna combined with the impedance matching model and studied the influence of each parameter on the antenna, which provides the direction for the optimization of the antenna design. The antenna model has relatively independent performance changes, so it is easy to adjust the parameters to make the antenna meet certain application requirements.

At the same time, combined with optical, electromagnetic, and scientific computing visualization technology, the electromagnetic field radiation intensity of the tag antenna is more intuitively expressed [9–11]. The image processing method is used to optimize the interference factors of the electromagnetic radiation field around the tag antenna, and the antenna is reverse designed. Therefore, the RFID tag is designed with minimal limits to optimize the spatial electromagnetic field strength and electromagnetic wave receiving capability. The research method in this paper provides a novel method for antenna design. The motivation of the paper is to simplify the complexity of the antenna design algorithm. Compared with the traditional method, the method reduces the calculation difficulty and simplifies the parameter selection of the antenna design [12–14].

The content of this study is as follows: The second part introduces the antenna structure, analyzes the influence of the main parameters of the antenna on the antenna performance, and introduces the reverse design method of the RFID antenna. The third part simulates and analyzes the test results of the influence of antenna parameters on performance, and inversely designs the antenna. The fourth part introduces the conclusion and significance of this chapter.

4.5.2 Design of UHF RFID Dipole Tag Antenna

(1) RFID antenna reverse design method

Figure 4.24 is a half-wave dipole antenna model with the impedance matching loop. For l is the length of the dipole antenna; a, b, w' is the length, width, and

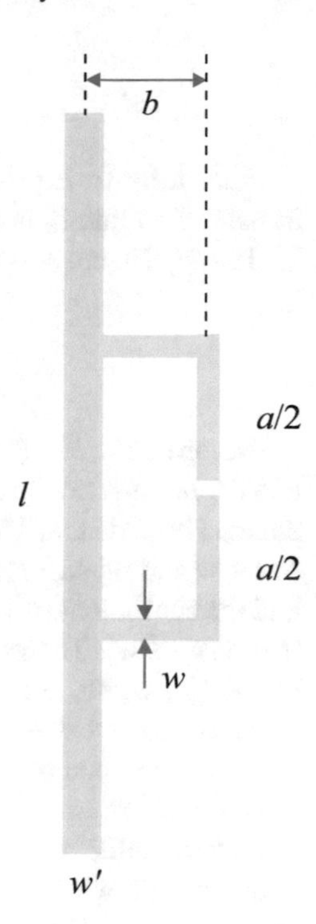

Fig. 4.24 Half-wave dipole antenna model

line width of the impedance matching loop; and w is the width of the dipole antenna.

The length of the half-wave dipole antenna is determined by Eq. (4.39):

$$L = \frac{\lambda}{2} = \frac{c}{2f\sqrt{\varepsilon}} \tag{4.39}$$

Among them, c is the propagation velocity of light in vacuum, f is the center frequency of the antenna, and ε is the relative dielectric constant.

The impedance value of the dipole antenna can be expressed as [15, 16]

$$Z_{in} = \frac{2Z_t(1+\alpha)^2 Z_A}{2Z_t + (1+\alpha)^2 Z_A} \tag{4.40}$$

$$Z_{in} = \frac{1}{\frac{1}{(1+\alpha)^2 Z_A} + \frac{1}{2Z_t}} \tag{4.41}$$

The size of antenna impedance is determined by $(1+\alpha)^2 Z_A$ and Z_t by (4.41). The impedance between the antenna and the chip is Z_A and the impedance between the impedance loop and the antenna is Z_t. The size is:

$$Z_t = jZ_0 \tan\left(\frac{ka}{2}\right) \tag{4.42}$$

According to Eq. (4.42), when the impedance loop length a is increased, the transmission line Z_t portion is increased.

The coefficient α is determined by (4.43):

$$\alpha = \frac{\ln b - \ln 8.25w}{\ln b - \ln 0.25w'} \tag{4.43}$$

According to Eq. (4.43), the matching impedance loop width b, the line width w, and the antenna line width w' all affect the coefficient α, and thus, they can change the antenna impedance. The influence of b on the coefficient is limited, and the change of the w and w' will also have a certain effect on α, but the coefficient of w' is 8.25 and the coefficient of w is 0.25, so the impact of w on antenna impedance is greater than that of w'. Therefore, the main parameters affecting impedance matching are the length, width, and line width of the impedance matching loop.

Since the half-wave dipole operates in the ultra high frequency and the total length of the tag antenna reaches 16 cm, this size of the antenna is very difficult to use in most scenarios. In order to reduce the size of the label while working in the ultra high frequency band, the dipole antenna is bent. According to Fig. 4.24, an antenna model diagram is drawn. The dipole antenna is bent and the left and right radiators are bent into three equal-length sections to obtain the antenna model shown in Fig. 4.25. The simulation experiment mainly studies the effect of antenna geometric parameters on the antenna impedance and antenna performance. The parameters of the bent dipole antenna are consistent with the half-wave dipole, except for the antenna arm length, as shown in Fig. 4.25.

(2) RFID antenna reverse design method

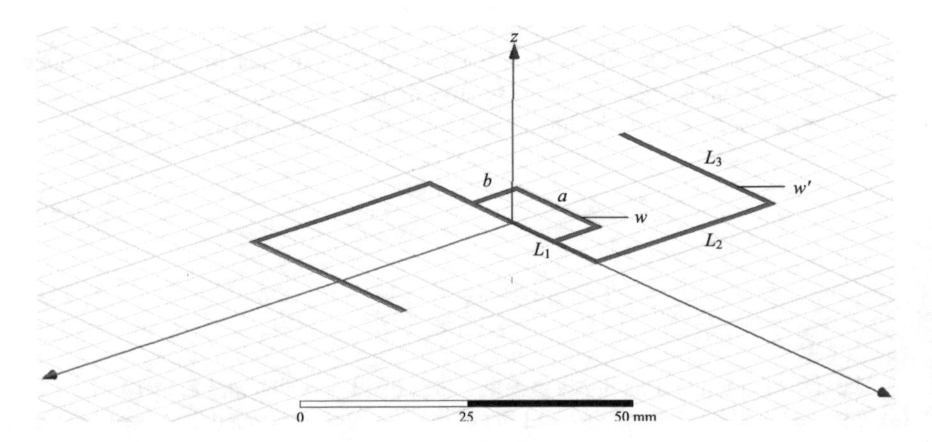

Fig. 4.25 Bent dipole tag antenna model

The specific absorption rate (SAR) describes the absorption characteristics of the medium to the electromagnetic field energy. SAR is defined as the rate of absorption of electromagnetic energy by a medium in the same environment. The medium can be regarded as a conductive medium with a dielectric constant of σ. When the effective amplitude of the electric field strength in the medium is E, the power loss value of the microwave in a unit volume of biological tissue can be expressed by Eq. (4.44):

$$P = \sigma \frac{E^2}{2} \tag{4.44}$$

The SAR image can visually display the intensity of the radiant energy of the antenna in a certain direction. Its intensity is related to antenna loss and gain. By observing and processing the antenna SAR image, the antenna tag can be reverse engineered to optimize the antenna radiation capability. By processing the SAR images, some important parameters of the radiation performance of the RFID antenna are obtained. These parameters can provide a basis for antenna design.

The general process of antenna SAR image processing is as follows. The SAR images of the tag antenna in a vacuum environment are extracted and binarized. Most of the original images are converted into binary information, thereby reducing unnecessary information in the image. The images are averagely filtered to eliminate image noise.

Among them, when the original images are relatively clear, the SAR images can be directly converted into a binary image, and a sharp contour can be formed after extracting the edge. However, some SAR images are generally not particularly clear. They need to be converted to a grayscale image, which is then filled with grayscale and then converted into a binary image.

Threshold segmentation is used for the preprocessed image. Determining a gray threshold T, the divided image can be expressed by

$$g(x, y) = \begin{cases} 1, & f(x, y) \geq T \\ 0, & f(x, y) < T \end{cases} \tag{4.45}$$

where $f(x, y)$ is the input image and $g(x, y)$ is the output image. If the pixel in $f(x, y)$ is greater than the set threshold, the region is the target image region. Otherwise, it belongs to the background area.

After the image is edge-detected, the feature value of the image is extracted. The eigenvalues can better express the main features and attributes of the target area. In this paper, the area, center of gravity, perimeter, and eccentricity are selected as the features of the image.

The image feature value is calculated as follows. Suppose the image size is $M \times N$, the coordinates of a certain point are (x, y), $f(x, y)$ represents the gray value of the SAR image at the coordinates, $O(x, y)$ represents the coordinate information located in the target area, $O(x, y) = \begin{cases} 1 \nu f(x, y) \in O \\ 0, \text{otherwise} \end{cases}$, and the area of the image area is calculated by Eq. (4.46):

$$A = \sum_{x=0}^{M-1} \sum_{y=0}^{N-1} O(x, y) \tag{4.46}$$

The calculation formula for the position of the center of gravity of the image is Eq. (4.47):

$$\begin{cases} \overline{x} = \dfrac{1}{A} \displaystyle\sum_{x=0}^{M-1} \sum_{y=0}^{N-1} xO(x, y) \\ \overline{y} = \dfrac{1}{A} \displaystyle\sum_{x=0}^{M-1} \sum_{y=0}^{N-1} yO(x, y) \end{cases} \tag{4.47}$$

The perimeter is represented by an eight-way chain code, which is a method of describing a curve or boundary using the coordinates of the starting point of the curve and the direction code of the boundary point. The eccentricity is the ratio of the distance between the focal points to the length of the long axis.

According to the requirements of the antenna performance in the experimental application, the selection range of the eigenvalues of the antenna SAR image can be determined, and the structural parameters of the antenna are reversely derived.

4.5.3 UHF RFID Tag Bend Dipole Antenna Analysis

The basic parameters of the bent dipole antenna are shown in Table 4.8. The parameters of the antenna include the length L_1, L_2, L_3 of the bent dipole; the length a, width b, and line width w of the impedance loop; and the antenna line width w'. We

Table 4.8 The bent dipole model antenna parameters

Parameters	Reference value/mm
L_1	20
L_2	35
L_3	25
w	1
a	20
b	10
w'	1

adjust a certain parameter of the antenna model and use HFSS simulation to analyze the change of antenna performance. We observe and analyze the specific effects of each parameter on antenna performance.

(1) Antenna performance change analysis

(1) Analysis of the influence of antenna length on antenna performance

When the length of the bent dipole antenna L_3 is 20, 25, 30, and 35 mm, the total length of the antenna is equivalent to 150, 160, 170, and 180 mm. As can be seen from Fig. 4.26, the center frequency of the bent dipole antenna is significantly shifted to the right compared to the basic half-wave dipole antenna. Taking a length L_3 of 35 mm as an example, the theoretical center frequency of the half-wave dipole antenna is 833 MHz, which is smaller than 920 MHz in the figure.

Because the antenna is bent, the mutual inductance between the antenna arms causes the antenna impedance to change. The antenna does not match the impedance of the chip at the original frequency, causing the center frequency of the antenna to shift. Since the UHF is mainly used in the 920–925 MHz band, the following studies take L_3 as 35 mm.

(2) Analysis of the influence of the impedance matching loop length a on antenna performance

As shown in Fig. 4.27, when the impedance matching loop route is continuously increased, the return loss value generally increases, and the center frequency shifts to the right. The standard impedance of the tag chip is 50. From Fig. 4.28a, b, it can be seen that when the antenna is operating at the center frequency, the impedance of the antenna is close to 50, and the antenna and chip impedance is matched. When

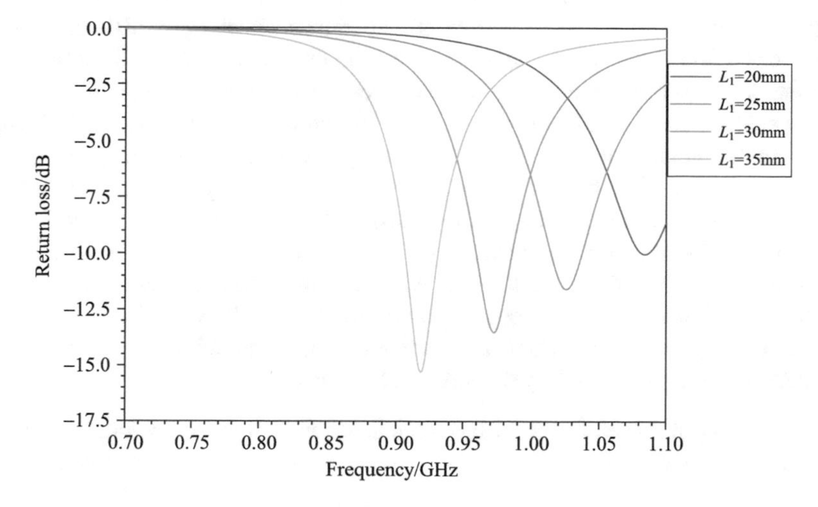

Fig. 4.26 Changes of the return loss of bent dipole antenna

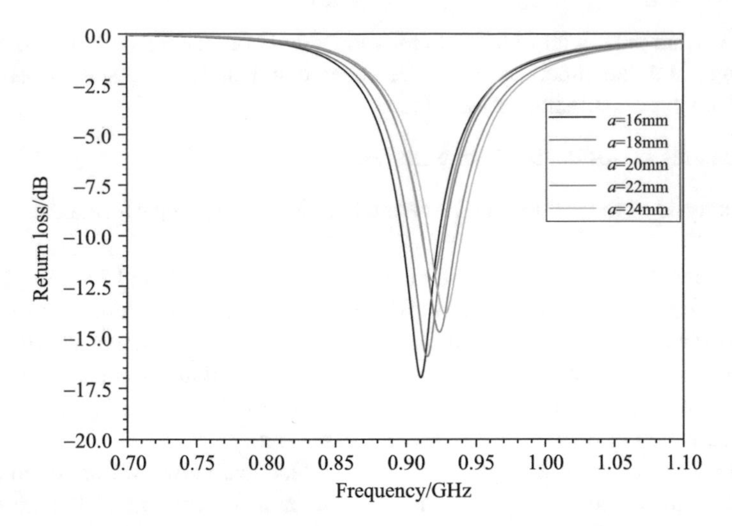

Fig. 4.27 Changes of return loss with impedance loop length

a is 16 mm, the antenna impedance is closest to 50, so the return loss is minimal at this time. As *a* increases, the real part of the impedance increases in turn, while the imaginary part gradually decreases. It can be inferred that when *a* is less than 16 mm, the antenna does not match the impedance of the chip, and the return loss will increase instead. Therefore, when the impedance matching loop length *a* is 16 mm or 18 mm, the antenna performance is superior.

As can be seen from Fig. 4.28, the values of the real and imaginary parts of the impedance change significantly. When the impedance matching loop length is determined, the antenna impedance curve will always approach the conjugate matching condition at a certain frequency, which is also the reason why the antenna center frequency shifts. However, the degree of impedance matching is different, so the return loss of the antenna does not increase as the length of the impedance matching loop increases. When *a* is 20 mm, the return loss value is relatively large.

The antenna gain visualization images are shown in Fig. 4.29. The red part represents a high antenna gain and the green part represents a low antenna gain. As can be seen from Fig. 4.29, the gain of the five images is basically the same. This shows that the antenna parameter changes the return loss, but has no effect on the antenna gain. In antenna design, when changing the impedance loop length, it is no longer necessary to consider the antenna gain. When changing the other parameters of the antenna impedance matching loop, the simulation results show that the antenna gain is also unchanged. This shows that the antenna gain is not affected by the antenna parameters, but it is affected by the antenna structure.

(3) Analysis of the influence of the impedance matching loop width on antenna performance

Changing the size of the impedance matching loop width *b*, the result is shown in Fig. 4.30. The center frequency of each antenna works slightly changed, but the

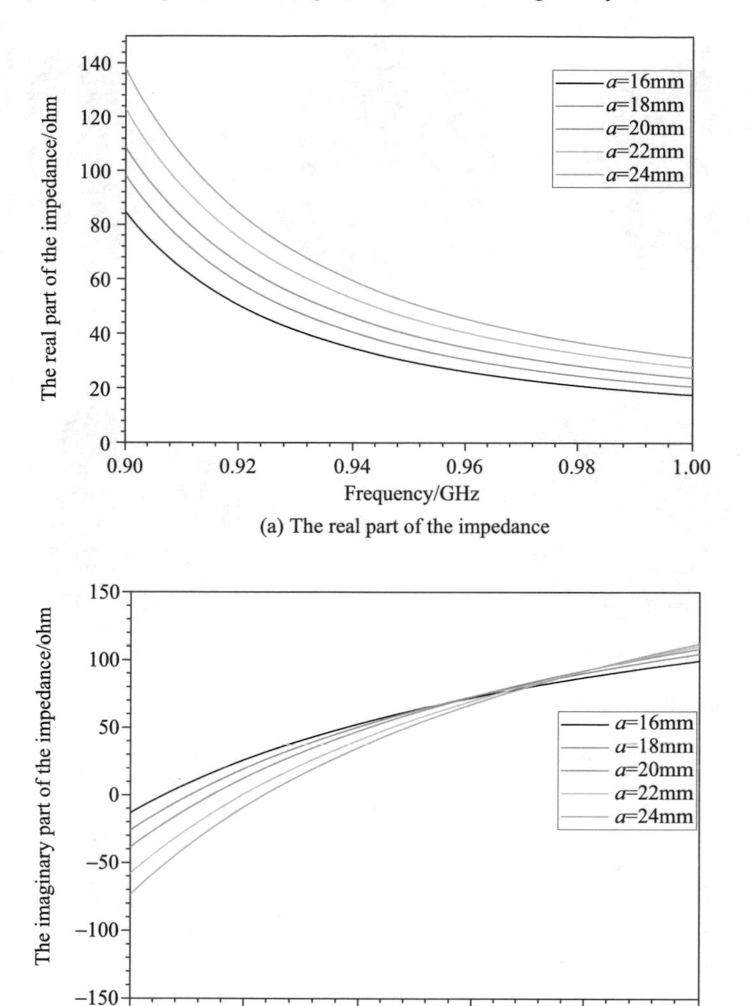

(a) The real part of the impedance

(b) The imaginary part of the impedance

Fig. 4.28 Changes of impedance with the impedance loop length

return loss changes. When b is 12 mm, the loss is minimal. When b changes, the loss slightly changes, but the center frequency is basically unchanged. The performance of the antenna changes relatively little.

(4) Analysis of the influence of the impedance matching loop line width w on antenna performance.

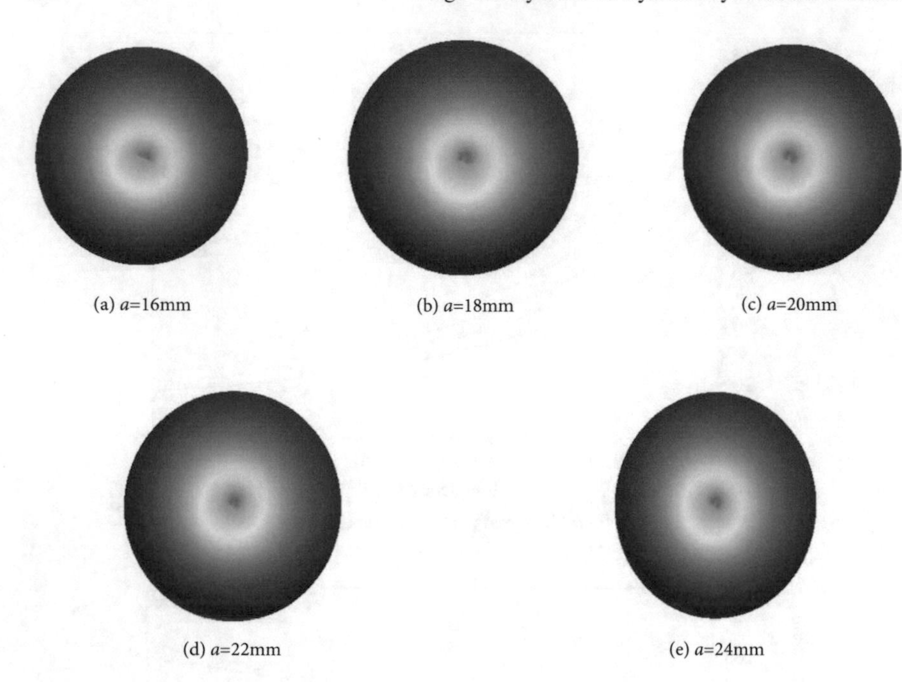

Fig. 4.29 Visualized images of antenna gain

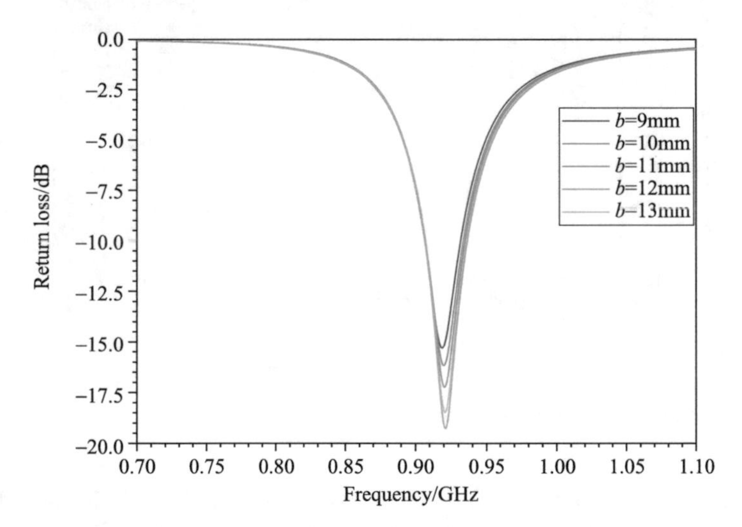

Fig. 4.30 Changes of return loss with impedance loop width

When the line width w of the impedance matching loop is increased from 0.8 mm to 1.6 mm, the center frequency of the antenna and return loss both increase as shown in Fig. 4.31. When w is 0.8 mm, the antenna and the chip achieve a good match.

(2) SAR image analysis

The antenna electromagnetic field radiation intensity images are represented in Fig. 4.32, where L_3 is 32,…, 35, 36 mm. Except for the antenna parameters mentioned, other antenna parameters are unchanged. The antenna center frequency selected for this simulation is 920 MHz. Among them, the closer the color is to the blue portion, the lower the specific absorption rate. It shows that the radiant energy intensity of the antenna in this direction is smaller.

The SAR images in Fig. 4.32 are binarized to obtain the image results in Fig. 4.33. The radiation capability of the antenna can be visually seen in Fig. 4.33. As can be seen from the figure, when $L_3 = 35$mm, the area of the image is large, that is, the antenna performance is excellent. Moreover, when the length of L_3 changes, the

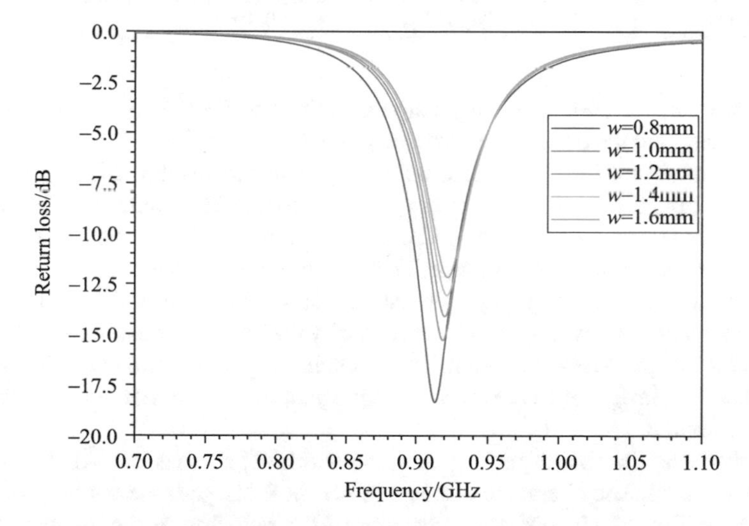

Fig. 4.31 Changes of return loss with impedance matching loop line width

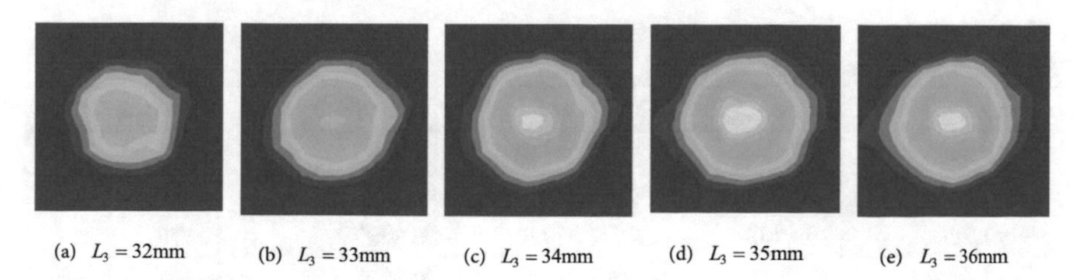

Fig. 4.32 SAR image

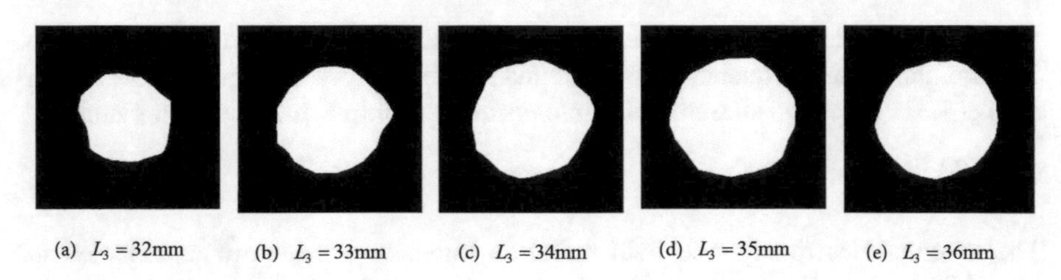

(a) $L_3 = 32$mm (b) $L_3 = 33$mm (c) $L_3 = 34$mm (d) $L_3 = 35$mm (e) $L_3 = 36$mm

Fig. 4.33 SAR binarized image

Table 4.9 SAR image feature values

Eigenvalues	Area	Centroid	Eccentricity	Perimeter
$L_3 = 32$mm	58,071	(271.9,282.3)	0.4282	894
$L_3 = 33$mm	81,661	(282.8,284.7)	0.4059	1041
$L_3 = 34$mm	96,767	(278.9,279.8)	0.3577	1136
$L_3 = 35$mm	104,257	(279.4,278.8)	0.2964	1180
$L_3 = 36$mm	96,020	(275.7,279.9)	0.3169	1123

performance of the antenna changes significantly. The feature values of the images are extracted, and the data is as shown in Table 4.9.

A certain rule can be obtained from the image and data results. The larger the area of the SAR image, the greater the energy radiated by the RFID antenna. The position of the center of gravity of the image can represent the position where the radiation energy of the antenna is the strongest. There is basically no change in the position of the center of gravity in the image, indicating that the direction of the radiant energy of the antenna does not change. The eccentricity and perimeter can reflect the shape of the SAR image. When the eccentricity is small, the graph is close to a circle. In the case where the image areas are substantially equal, the pattern of the long perimeter is more irregular.

By changing the size of the antenna line width w', a series of SAR images can also be obtained. The characteristic data of the SAR image is shown in Table 4.10. The area in Table 4.10 shows the intensity of the radiation of the antenna on that face. Compared to Table 4.9, the change in area is not very obvious. This shows that

Table 4.10 SAR image feature values

Eigenvalues	Area	Centroid	Eccentricity	Perimeter
$w' = 0.8$mm	100,412	(273.3,281.8)	0.3390	1153
$w' = 1.0$mm	104,257	(279.4,278.8)	0.2964	1180
$w' = 1.2$mm	103,448	(273.2,279.6)	0.3563	1178
$w' = 1.4$mm	102,356	(275.8,280.8)	0.3805	1168
$w' = 1.4$mm	101,344	(271.5,284.1)	0.3837	1154

the arm length has an effect on the antenna performance higher than the antenna line width. The center position changes within a certain range, but the change is not large. The radiation direction of the antenna is basically unchanged.

The length of the antenna bend may affect the performance of the antenna. In the above, when $L_2 + L_3 = 70$ mm, the radiation of the antenna is the strongest. Different SAR images can be obtained by guaranteeing $L_2 + L_3 = 70$ mm and changing the lengths of L_2 and L_3. After image processing, the extracted features are shown in Table 4.11. The antenna performance is different under the different bending lengths of the antenna. When $L_2 = 35$ mm, $L_3 = 35$ mm, the SAR image area of the antenna is the largest, indicating that the radiation capability of the antenna is the strongest currently. The antenna losses in the five cases were –7.6801 dB, –11.2734 dB, –10.4744 dB, –8.0956 dB, and –6.2453 dB, respectively. The ratio of antenna return loss is 0.68:1:0.93:0.72:0.55. At the same time, the SAR image area ratio is 0.82:1:0.87:0.70. The ratios of the two are relatively similar, indicating that the SAR image can accurately reflect the antenna radiation.

The bending length of the single arm on the antenna is changed, and the obtained antenna characteristic data is shown in Table 4.12. Not only the area of the image changes significantly, but also the center position changes, which gradually moves in one direction. This shows that the radiation direction of the antenna is also changing.

The position of the impedance matching loop is shifted, and the data of the SAR image feature values are shown in Table 4.13. The antenna position does not move, and only the impedance matching loop is translated. When the impedance loop moves in the positive direction of the *y*-axis, its moving distance is a positive number. The data shows that there is no obvious trend in SAR changes, and the difference in the changes is not very large. This shows that the main influencing factor of SAR is

Table 4.11 SAR image feature value of L_2 and L_3

Eigenvalues	Area	Centroid	Eccentricity	Perimeter
$L_2 = 30$mm, $L_3 = 40$mm	85,881	(275.5,282.7)	0.4360	1076
$L_2 = 35$mm, $L_3 = 35$mm	104,257	(279.4,278.8)	0.2964	1180
$L_2 = 40$mm, $L_3 = 30$mm	90,306	(273.2,284.4)	0.2875	1087
$L_2 = 45$mm, $L_3 = 25$mm	73,493	(274.3,279.0)	0.3551	979
$L_2 = 50$mm, $L_3 = 20$mm	63,339	(270.4,281.6)	0.4338	918

Table 4.12 SAR image feature values of single arm

Eigenvalues	Area	Centroid	Eccentricity	Perimeter
$L_2 = 20$mm, $L_3 = 50$mm	84,596	(278.4,295.6)	0.3771	1071
$L_2 = 30$mm, $L_3 = 40$mm	103,272	(274.7,286.3)	0.3676	1185
$L_2 = 40$mm, $L_3 = 30$mm	100,909	(273.3,281.2)	0.2743	1164
$L_2 = 50$mm, $L_3 = 20$mm	85,082	(266.7,272.3)	0.3082	1068
$L_2 = 60$mm, $L_3 = 10$mm	69,099	(264.1,271.1)	0.4117	956

Table 4.13 SAR image features at different loop positions

Moving distance	Area	Centroid	Eccentricity	Perimeter
4mm	98,954	(269.6,279.2)	0.2901	1147
2mm	99,722	(270.6,280.2)	0.3364	1149
0	104,257	(279.4,278.8)	0.2964	1180
−2mm	106,675	(272.9,281.7)	0.4009	1201
−4mm	102,201	(272.9,283.2)	0.3025	1177

the length of the antenna arm, and the impedance loop does not affect the radiation performance.

(3) Contrast and discussion

From the above simulation results, the center frequency of the bent dipole antenna is mainly determined by the total arm length of the antenna, but it deviates from the basic half-wave dipole antenna theory. Due to the mutual inductance between the antenna arms, the impedance variation of the bent dipole antenna is not the same as the theoretical model, but the influence degree of the parameters estimated from the antenna theory on the impedance is consistent with the simulation result. Changing the impedance loop length can significantly change the center frequency and return loss of the antenna. Relatively speaking, the width and line width of the impedance loop have less effect on the performance of the antenna. At the same time, when the antenna parameters are changed, the antenna gain is substantially unchanged. Only one of the real or imaginary parts of the antenna impedance changes, and the center frequency of the antenna generally does not change much.

Through the calculation of SAR image features, we can get that when the antenna parameters satisfy $L_2 = 35$ mm, $L_3 = 35$ mm, $w' = 1$ mm, $w = 0.8$ mm, the performance of the antenna is the best. Several features of the antenna SAR image can clearly reflect some of the performance of the antenna. The area of the image can reflect the radiation intensity of the antenna, which is related to the loss and gain of the antenna. The radiation intensity is related to the antenna signal propagation distance, so the antenna SAR image area can intuitively reflect the antenna transmission distance. The center position of the SAR image can reflect the main radiation direction of the antenna. The eccentricity and perimeter can reflect the radiation range of the antenna. In general, the smaller the eccentricity and perimeter, the easier it is for the antenna to cover the signal in one direction. The simulation results show that the SAR image analysis results can correctly reflect the antenna performance.

At present, the methods of antenna optimization design mainly include electromagnetic simulation calculation methods [11, 12] and neural network optimization methods [13]. In the electromagnetic calculation method, it is necessary to compare the effects of different antenna parameters on the antenna. Then, modify the antenna parameters one by one to make the antenna meet the design requirements in a certain frequency band. Electromagnetic calculation methods require iterative calculations and take a long time. Since the performance is affected by different parameters, the adjustment parameters are more complicated and the calculation workload is huge. Compared with other electromagnetic calculation methods, this method can acquire a large number of images in a short time. The imaging method can quickly determine the antenna parameters, and the real-time performance is good. Under the same antenna structure, the antenna parameters can be changed according to the design requirements without more calculations. The neural network optimization method has high precision. However, it is a complicated job, and some parameters of the neural network need to be adjusted according to different prediction work.

The antenna design process is relatively simple by the antenna reverse design method of SAR image analysis. Moreover, the radiation intensity and distribution information of the antenna can be discriminated by image information and feature values, and the result is a more intuitive and real-time result.

4.6 Conclusion

Firstly, this chapter presents the research of RFID tag positioning method based on three-dimensional space multi-tag positioning system. In this chapter, the horizontal and vertical CCD are used to collect relative experimental images. Threshold segmentation and edge detection algorithms are used to draw the bottom pattern of the label holder ($r_n \cos\theta_n$, $r_n \sin\theta_n$) taken by the vertical CCD camera. Then used the edge detection algorithm to match the area where the label is taken by the horizontal CCD camera to determine its height information h_n. That is, the complete coordinate information of the obtained labels is distributed in three-dimensional space. Therefore, the problem of accurate identification of the spatial distribution of RFID multi-tags in a complex environment is obtained, which provides an important reference for improving the research of RFID tag reading performance.

Then, according to the practical application requirements of the UHF antenna, a bent dipole antenna is designed and combined with an impedance matching loop. The structure of the antenna is simple, and the structural parameters are easy to adjust to achieve a good impedance matching effect between the antenna and the chip. On this basis, through the analysis of SAR images, the antenna can be reverse engineered according to the actual application requirements of the antenna, so that the antenna design can meet the application requirements. The reverse design method is simple and more intuitive to reflect the gap between design and application. This paper provides guidance and direction for the design of RFID antennas.

References

1. Yu X, Wang D, Zhao Z (2019) Semi-physical verification technology for dynamic performance of internet of things system. Springer, Singapore
2. Liu J, Bu F (2019) Improved RANSAC features image-matching method based on SURF. J Eng 2019(23):9118–9122
3. Faudzi NM, Ali MT, Ismail I et al (2014) Compact microstrip patch UHF-RFID tag antenna for metal object. In: IEEE symposium on wireless technology and applications. Kota Kinabalu, Malaysia, pp 160–164
4. Marrocco G (2008) The art of UHF RFID antenna design: Impedance-matching and size-reduction techniques. IEEE Antennas Propag Mag 50:66–79
5. Zuffanelli S, Zamora G, Aguila P et al (2016) Analysis of the split ring resonator (SRR) antenna applied to passive UHF-RFID tag design. IEEE Trans Antennas Propag 64:856–864
6. Luk WT, Yung KN (2008) Bending dipole design of passive UHF RFID tag antenna for CD, DVD discs. In: Asia-pacific microwave conference. Macau, China, pp 1–4
7. Yao Y, Cui C, Yu J et al (2017) A meander line UHF RFID reader antenna for near-field applications. IEEE Trans Antennas Propag 65:82–91
8. Qin C, Mo L, Zhou H, et al (2013) Dual-dipole UHF RFID tag antenna with quasi-isotropic patterns based on four-axis reflection symmetry. Int J Antennas Propag
9. Yang ES, Son HW (2016) Dual-polarised metal-mountable UHF RFID tag antenna for polarisation diversity. Electron Lett 52:496–498
10. Hamani A, Yagoub MC, Vuong TP et al (2016) A novel broadband antenna design for UHF RFID tags on metallic surface environments. IEEE Antennas Wirel Propag Lett 16:91–94
11. Coisson M, Barrera G, Celegato F et al (2017) Hysteresis losses and specific absorption rate measurements in magnetic nanoparticles for hyperthermia applications. Biochem Biophys Acta 1861(6):1545–1558
12. Coïsson M, Barrera G, Celegato F et al (2016) Specific absorption rate determination of magnetic nanoparticles through hyperthermia measurements in non-adiabatic conditions. J Magn Magn Mater 415:2–7
13. Golestanirad L, Rahsepar AA, Kirsch JE et al (2019) Changes in the specific absorption rate (SAR) of radiofrequency energy in patients with retained cardiac leads during MRI at 1.5T and 3T. Magn Resonance Med 81(1): 653–669
14. Li Y, Yang X, Yang Q et al (2011) Compact coplanar waveguide fed ultra wideband antenna with a notch band characteristic. AEU-Int J Electron Commun 65:961–966
15. Nassar IT, Weller TM (2015) A novel method for improving antipodal Vivaldi antenna performance. IEEE Trans Antennas Propag 63:3321–3324
16. Xiao LY, Shao W, Liang TL et al (2017) Artificial neural network with data mining techniques for antenna design. In: IEEE international symposium on antennas and propagation and USNC/URSI national radio science meeting. San Diego, CA, USA, pp 159–160
17. Balanis CA (2016) Antenna theory: analysis and design. Wiley, New York

Chapter 5
Optimization Algorithm and RFID System Physical Anti-Collision

Radio-frequency identification (RFID) technology is a wireless communication technology that enables users to uniquely identify tagged objects or people [1–4]. The passive RFID system, due to no direct power supply, is widely used in the field of warehouse management, smart transportation, healthcare industries, and so on. An important advantage of RFID technology is multi-target recognition at the same time, but the problem of improving the reading performance emergences. Nowadays, the common method to the problem is anti-collision algorithms, which solve the data conflict of multiple tags within the same radio frequency (RF) channel, such as ALOHA algorithm and binary tree algorithm [5–8]. However, these algorithms could not improve the reading performance to some extent in practice.

In recent years, the application of the camera on the target measurement has been widely used [9]. In [10], Ma presented a line-scan CCD camera calibration method in 2D coordinate measurement. In [11], Dong used single linear array CCD to measure the vertical target density. In [12], Fahringer described a novel 3D, three-component (3C) particle image velocimetry technique which was based on volume illumination and light field imaging with a single camera. Zhou presented a novel model and the corresponding calibration approach which took the sensors as an integrated structure with the viewpoint [13]. Chen proposed a novel non-contact, full-field, 3D, multi-camera digital image correlation (DIC) measurement system. In the proposed system, multiple cameras are combined as a single system [14]. Venkataraman used the images captured by camera arrays to measure the depth [15]. However, comparing with the two CCD cameras, using a single CCD camera to measure the target 3D coordinates needs to constantly adjust the camera position to obtain the same state of the object image from different angles. The operation is complex, real-time poor and it is difficult to adapt to the requirements of modern warehousing logistics. So, in this paper, two CCD cameras (vertical and horizontal cameras) are used to measure the 3D coordinates of the RFID tags.

In the field of nonlinear modeling, neural networks have many advantages. The neural networks do not need to construct mathematical function-based models. It can approximate any nonlinear function or the nonlinear function relations of complex

X. Yu et al., *Physical Anti-Collision in RFID Systems*,
https://doi.org/10.1007/978-981-16-0835-3_5

multiple inputs and outputs. Besides that the neural network has a strong adaptive and self-learning ability. So, the neural network has been widely used in many situations. In [16], Wang built a wind power range prediction model based on the multiple output property of backpropagation (BP) neural network. Then, the improved particle swarm optimization (PSO) algorithm was used to optimize the model. In [17], Ding proposed the genetic algorithm (GA) to optimize the BP algorithm. The genetic algorithm can overcome BP's disadvantage of being easily stuck in a local minimum. In [18], Wang presented a two-layer decomposition technique and then developed a hybrid model based on fast ensemble empirical mode decomposition (FEEMD), variational mode decomposition (VMD), and BP neural network. In [19], Doucoure developed a prediction method for renewable energy sources to achieve an intelligent management of a microgrid system. The proposed method was based on the multi-resolution analysis of the time-series by means of wavelet decomposition and artificial neural networks. Sharma proposed a mixed wavelet neural network (WNN) for short-term solar irradiance forecasting, with initial application in tropical Singapore [20].

In order to optimize the reading performance of tags, this chapter mainly models the nonlinear relationship between the coordinate data of multi-tag and the reading distance through the following three aspects, and reads the optimal geometric distribution of tags. Therefore, the main contents of this chapter are as follows. Section 5.1 introduces the multi-tag optimization method based on particle swarm optimization (PSO). Section 5.2 introduces the multi-tag anti-collision optimization algorithm based on support vector machine (SVM). Section 5.3 introduces the multi-tag anti-collision optimization algorithm based on wavelet. Finally, conclusion is given.

5.1 Physical Anti-Collision Based on Particle Swarm Optimization (PSO)

5.1.1 Design and Application of Detection System

(1) Structure of detection system

In order to simulate the mobile product and the environment, a RFID detection system has been designed (shown in Figs. 5.1 and 5.2). The RFID detection system mainly consists of three parts, including the acquisition system, detection system, and control system. The acquisition system is composed of a vertical CCD, a horizontal CCD, an servo motor and a laser ranging sensor. The detection system includes a reader, a certain number of antennas, and an antenna frame. The control system consists of a pallet, a transportation device, and a control computer. The application inspection items of the system include the test of tags' reading range, anti-collision performance, and location optimization.

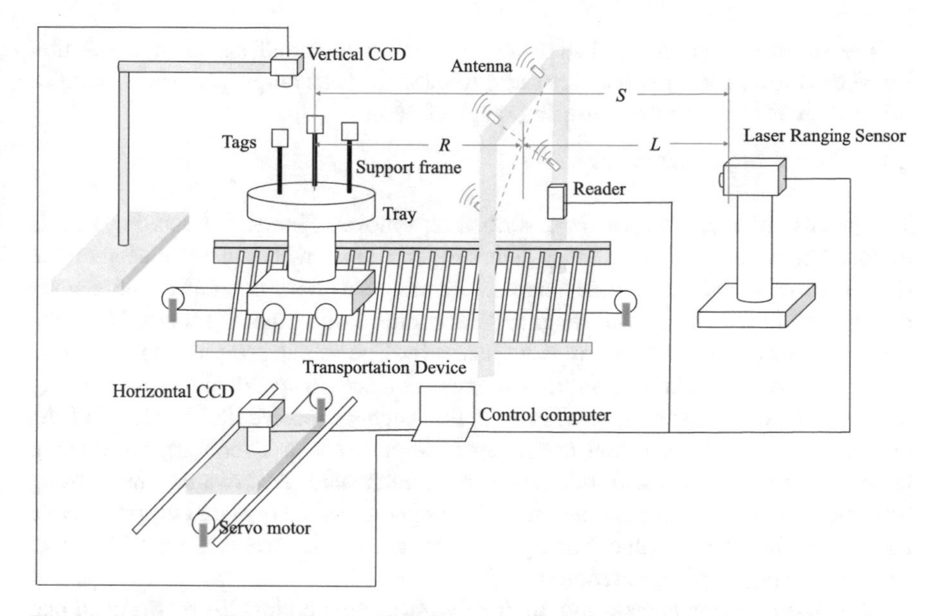

Fig. 5.1 Schematic diagram of dual CCD detection system

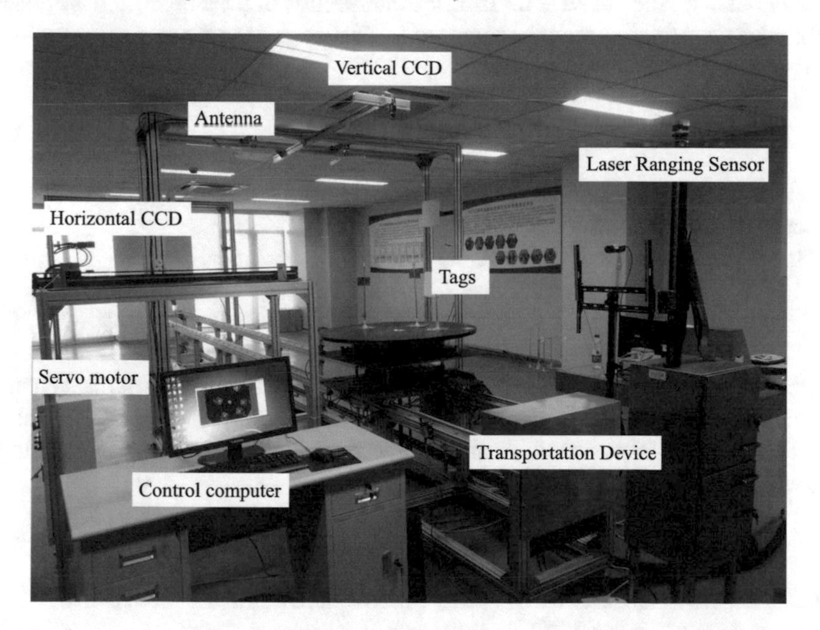

Fig. 5.2 Physical diagram of dual CCD detection system

The tags are the commercial UHF tags, the readers are Impinj speedway revolution R420, and the readers' antennas are Larid A9028. The laser ranging sensor is Wenglor X1TA101MHT88 with its measuring range of 50 m.

(2) Process of measurement

The process of measurement is described as follows. Firstly, the rotating tray is mounted on a transportation device. The support frames with tags are placed on the rotating tray. A reader and some antennas are mounted on the antenna frame. Subsequently, the beam of the laser ranging sensor aims at the support frames. Next, the cycle number of the rotating tray is set manually. The rotating tray which transports tags simulates products in and out warehouse at a certain speed. When the rotating tray with the tags enters the reading area, the antennas receive the RF signal of the tags, and then send the signal to the laser ranging sensor. Therefore, the reading range between the tags and antennas can be calculated. Afterwards, the rotating tray returns to the initial point and the above operations are repeated until the cycle number reaches the set value. Finally, the average value is used as the reading range between the tags and the antennas [21, 22].

The reading range is measured indirectly. Adjust the optical lift platform so that the laser beam can be aimed at the box; the intersection of laser beam and the antenna frame plane is defined as the reference points. The range from the tags to the reference point is

$$R = S - L \tag{5.1}$$

where S is the range of laser ranging sensor to tags and L is the range of laser ranging sensor to reference point.

The range between tag and i_{th} RFID antenna is

$$R_i = (R^2 + H_i^2)^{1/2} \tag{5.2}$$

where H_i is the distance between i_{th} RFID antenna and reference point. When the antennas are fixed on the antenna frame, the range between reference point and i_{th} RFID antenna is measured manually and then inputted to the program. In the measurement, the tags are distributed randomly in space.

5.1.2 Establishment of 3D Tag Network Based on Template Matching and Edge Detection

The tag network consists of multiple nodes, where each node represents a specific location for the corresponding tag. In this experiment, the location of the tag is distributed in space randomly, as shown in Fig. 5.3.

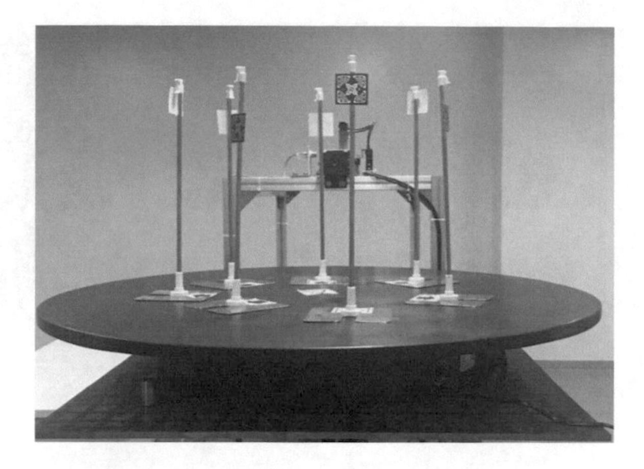

Fig. 5.3 The spatial distribution of multi-tag

(1) Measurement of horizontal coordinate

In order to calculate the spatial coordinates of tags, template matching method has been used. Template matching is to compare the source image with the template, usually a smaller image, so that the same or similar region in the source image can be found and extracted. The similarity between the image and the template is judged by the correlation coefficient.

Suppose $f(x, y)$ is the original image of A × B, and $t(x, y)$ is the template image of J × K (J ≤ A, K ≤ B), then the correlation coefficient is defined as

$$D(x, y) = 2\sum_{j=0}^{J-1}\sum_{k=0}^{K-1}[f(x + j, y + k) \cdot t(j, k)] \tag{5.3}$$

when $D(x, y)$ reaches the maximum value, the template corresponds to the corresponding region in the original image.

We marked the base of each tag frame to facilitate the template matching, the results are shown in Fig. 5.4, where (a) shows the mark (template), (b) shows the matching results. As it can be seen from the figures, each tag is accurately matched and marked (Fig 5.5).

Thereafter, all the RFID tags are numbered. The distance r_i between the i_{th} RFID tag and the center of the turntable is measured by the method below

$$r_i = D \times (k_i/d) \tag{5.4}$$

where k_i is the distance between the i_{th} RFID tag and the center of the turntable in picture, D is the real radius of the turntable, d is the radius of the turntable in picture. D and d are measured before. The vertical CCD is fixed on the frame so that the ratio of D to d will be fixed during the measurement.

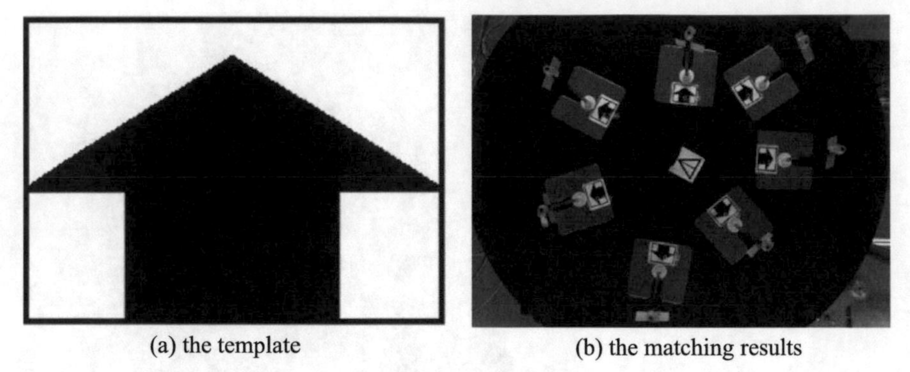

(a) the template (b) the matching results

Fig. 5.4 Template matching in horizontal direction

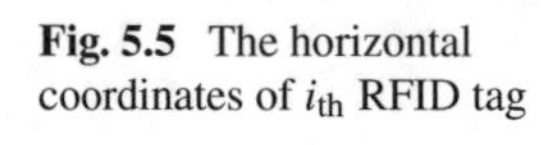

Fig. 5.5 The horizontal coordinates of i_{th} RFID tag

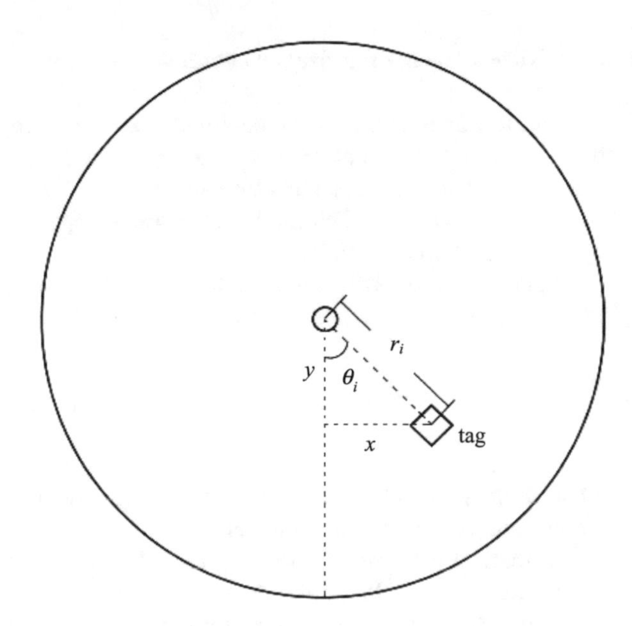

By adjusting the servo motor to drive the turntable, the i_{th} RFID tag will be rotated at a certain angle θ_i. Therefore, the horizontal coordinates of the i_{th} RFID tag is $(r_i \sin\theta_i, r_i \cos\theta_i)$.

(2) Measurement of vertical coordinate

During the measurement of the vertical coordinate of the tag, the horizontal CCD will automatically focus on the tag according to its horizontal coordinates. Firstly, the servomotor drives the tray to rotate so that the horizontal camera can face the i_{th} RFID tag. Assuming that the distance between the horizontal CCD and the center of the tray is L, the distance between the tag and the horizontal CCD should be as follows:

$$d_i = L - r_i \tag{5.5}$$

Afterwards, the distance that the horizontal camera can exactly focus on the i_{th} RFID tag should be calculated

$$l_i = \frac{fl'}{l' - f} \tag{5.6}$$

Where l' is the distance between the lens and CCD sensor inside horizontal camera, f is the focal length of the horizontal camera.

Finally, the distance needed to be adjusted for the horizontal camera is as follows:

$$\Delta L_i = d_i - l_i \tag{5.7}$$

If ΔL_i is larger than zero, the horizontal camera approaches the tag. Otherwise, the horizontal camera moves away from the tag.

Canny edge detection algorithm is used to detect the edge of the RFID tag and its frame [23–25]. First, the Gauss filter is used to get a smooth image

$$\begin{cases} G(x, y) = f(x, y) * H(x, y) \\ H(x, y) = \frac{1}{2\pi\sigma^2} \exp\left(-\frac{x^2+y^2}{2\sigma^2}\right) \end{cases} \tag{5.8}$$

where (x, y) is the coordinates of the pixels in the image, * represents the convolution, σ is the scale parameter, which determines the degree of smoothing of the filtering window on the image, $f(x, y)$ is the input image.

Then, the gradient and its direction of the pixel are calculated by the finite difference of the first-order derivative

$$\begin{cases} G(x, y) = \sqrt{G_h^2(x, y) + G_v^2(x, y)} \\ \theta(x, y) = \arctan \frac{G_h(x,y)}{G_v(x,y)} \end{cases} \tag{5.9}$$

where $G_h(x, y)$ and $G_v(x, y)$ are the partial derivative in the parallel direction and the vertical of the pixel (x, y) respectively. The formulas are as follows:

$$\begin{cases} G_h(x, y) = \frac{[I(x,y+1)-I(x,y)+I(x+1,y+1)-I(x+1,y)]}{2} \\ G_v(x, y) = \frac{[I(x,y)-I(x+1,y)+I(x,y+1)-I(x+1,y+1)]}{2} \end{cases} \tag{5.10}$$

where $I(x, y)$ is the gray value of the pixel.

The non-maximal value of the gradient magnitude $G(x, y)$ is suppressed, and all the pixels of the gradient magnitude $G(x, y)$ are linearly interpolated. At each pixel point, the gradient amplitude of center pixel of the neighborhood is compared with the linear interpolation result. If the gradient amplitude of the neighborhood center point is larger than the linear interpolation result, the pixel point is the edge point.

The double threshold algorithm is used to detect and connect the edge. First, the low threshold δ_1 and high threshold δ_2 should be set. Then, the image is divided into low threshold edge image $T_1[x, y]$ and high threshold edge image $T_2[x, y]$ according to these two thresholds. The edge is connected into a contour in the image $T_2[x, y]$. When the contour endpoint in $T_2[x, y]$ is reached, the corresponding neighborhood position of the low threshold edge image $T_1[x, y]$ is searched for an edge $T_3[x, y]$ that can be connected to the contour. If $T_3[x, y]$ and $T_2[x, y]$ can be connected to a curve or a straight line, it is retained, otherwise discarded. The edge $T_3[x, y]$ in $T_1[x, y]$ is collected until $T_3[x, y]$ and $T_2[x, y]$ are connected to a full edge.

To obtain the vertical position of the tag, firstly we should use template matching method to mark out each tag, as shown in Fig. 5.6. In Fig. 5.6, (a) represents the template, and (b)–(h) represent the matching results. The horizontal camera should be adjusted to focus on the i_{th} RFID tag so that the image of the i_{th} RFID tag can be obtained clearly. The edge of the i_{th} RFID tag and its frame can be obtained by edge detection method, and the number of pixels on the width of the i_{th} RFID tag and its frame can also be calculated. Afterwards, the vertical coordinate H_i can be calculated by the ratio method.

$$H_i = a \times (n_i / m_i) \tag{5.11}$$

where a is the real width of the RFID tag, n_i is the sum of the pixels on the width of the i_{th} RFID tag and its frame, m_i is the pixels on the i_{th} RFID tag. Therefore, the spatial coordinate of the i_{th} RFID tag is $(r_i \sin\theta_i, r_i \cos\theta_i, H_i)$.

A larger amount of data has been measured for different spatial distribution of the tags in this experiment. The results are shown in columns 1 to 21 in Table 5.1, and (x_i, y_i, z_i) represents the spatial position of the i_{th} tag. A set of data is selected randomly to plot the spatial structure of the RFID tag network. The result is shown in Fig. 5.7.

5.1.3 Prediction of Tag Distribution Based on PSO Neural Network

(1) Algorithm description

PSO algorithm has a very excellent performance in the multi-objective optimization [26, 27]. Therefore, PSO is introduced for optimization of multi-tag network, which is conducive to improve the reading performance of RFID system.

The PSO algorithm is a process that uses an individual (also called a particle) to move in the solution space to find the optimal solution [28]. Each particle is described by three parameters: velocity, position, and fitness. The velocity determines the direction and distance of the particle and adjusts itself with the influence of other particles. The fitness determines whether the particle is good or bad. The better the

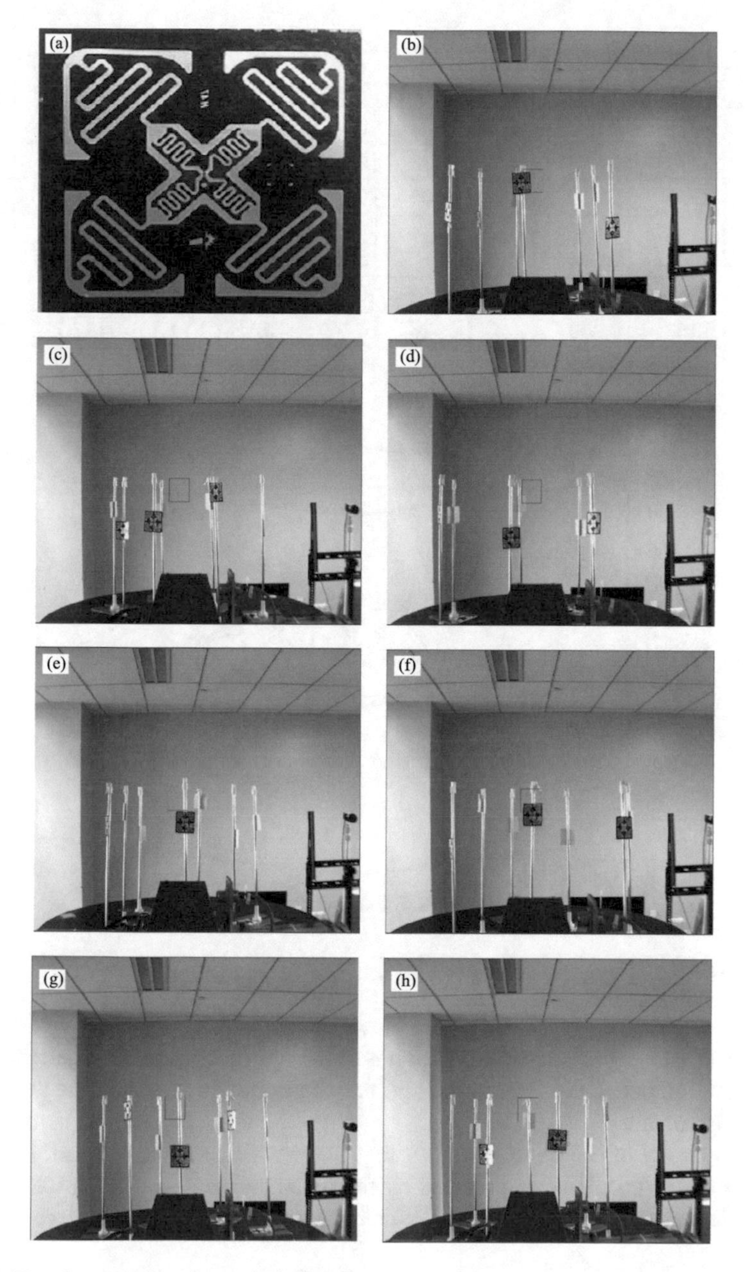

Fig. 5.6 Template matching in vertical direction

Table 5.1 The training data and results of PSO neural network

x_1/m	y_1/m	z_1/m	...	x_7/m	y_7/m	z_7/m	d_r/m	d_p/m	E/%
0.815	0.278	0.216	...	0.957	0.592	0.235	1.23	1.22	0.81
0.906	0.547	0.189	...	0.485	0.759	0.191	1.65	1.67	1.21
0.127	0.957	0.117	...	0.800	0.655	0.170	0.96	0.95	1.04
⋮	⋮	⋮		⋮	⋮	⋮	⋮	⋮	⋮
0.913	0.962	0.165	...	0.142	0.135	0.065	1.11	1.11	0
0.632	0.157	0.203	...	0.426	0.849	0.158	1.36	1.37	0.74
0.097	0.970	0.079	...	0.915	0.633	0.149	1.68	1.69	0.6

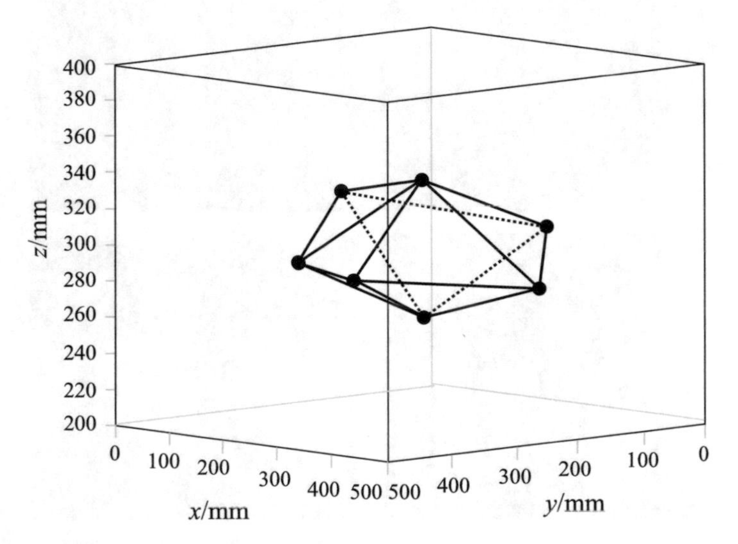

Fig. 5.7 The spatial structure of the RFID tag network

fitness is, the closer the particle position is to the optimal position. When the fitness is optimal, the particles move to the optimal position, which represents the optimal solution is solved.

The diversity of the population is a key factor which affects the performance of the PSO algorithm. It is represented by the distance between the particle and the geometric center of the population. We define the population geometric center $X'(t)$ as

$$X'(t) = \frac{1}{MN}\sum_{j=1}^{M}\sum_{i=1}^{N} X_{ij}(t) \tag{5.12}$$

where $X(t)$ is the particle position at time t, M is the dimension of solution space, N is the population size. Therefore, the population diversity can be expressed as follows:

$$I(X(t)) = \frac{1}{MN}\sum_{j=1}^{M}\sum_{i=1}^{N}[X_{ij}(t) - X'(t)]^2 \tag{5.13}$$

There are two main factors affecting population diversity, namely, inertia weight w and acceleration coefficients c_1, c_2. w, c_1, and c_2 all have the corresponding threshold (w', c_1', and c_2'), and the magnitude of these thresholds entirely depends on the parameters of particle and the extremes of the individual and the population [29]. When w, c_1, and c_2 are larger than their corresponding thresholds, the population diversity increases with the increase of these three coefficients. When w, c_1, and c_2 are smaller than their corresponding thresholds, the population diversity decreases with the increase of these three coefficients.

The sum of the error between the actual output and the predicted output is used as the fitness function F, the formula is as follows:

$$F = k\left(\sum_{1}^{n} abs(y_i - o_i)\right) \tag{5.14}$$

where n is the number of output nodes, y_i is the actual output of the i_{th} node, o_i is the prediction output of the i_{th} node, k is the coefficient.

Since the fitness is the sum of error, its value should be as small as possible. Therefore, the problem is transformed into the solution of the minimum value. Assuming that the solution space dimension of the solution problem is M, the velocity and position of the particle at time t can be expressed by the following vector

$$\begin{cases} \boldsymbol{V}(t) = (V_1(t), V_2(t), \cdots V_M(t)) \\ \boldsymbol{X}(t) = (X_1(t), X_2(t), \cdots X_M(t)) \end{cases} \tag{5.15}$$

The optimal position of the particle itself (i. e., the optimal position of the individual) can be expressed as

$$\boldsymbol{P}(t) = (P_1(t), P_2(t), \cdots P_M(t)) \tag{5.16}$$

The optimal location of population can be expressed as

$$\boldsymbol{G}(t) = (G_1(t), G_2(t), \cdots G_M(t)) \tag{5.17}$$

The particle moves in the solution space to find the optimal solution, and its parameters are updated with the time (iteration number), the updated formulas are as follows:

$$\begin{cases} V_m(t+1) = wV_m(t) + c_1 r(t)(P_m(t) - X_m(t)) + c_2 r'(t)(G_d(t) - X_m(t)) \\ X_m(t+1) = X_m(t) + V_m(t+1) \end{cases} \tag{5.18}$$

where $r(t)$ and $r'(t)$ are random numbers uniformly distributed over [0, 1] varying with time t. Considering the influence of the inertia weight w and the acceleration coefficients c_1, c_2 on the diversity of the population, the linear inertia weight and the time-varying acceleration coefficients are used in this paper. This method focuses on self-learning of $\boldsymbol{P}(t)$ and $\boldsymbol{G}(t)$ in the iterative process, which is beneficial to improve the overall performance of the algorithm. The corresponding formulas are shown as follows:

$$\begin{cases} w(t) = w_e + (w_i - w_e)(t\text{max} - t)/t\text{max} \\ c_1(t) = c_{11} + (c_{12} - c_{11})t/t\text{max} \\ c_2(t) = c_{21} + (c_{22} - c_{21})t/t\text{max} \end{cases} \tag{5.19}$$

where w_i and w_e are the initial and final values of the inertia weights; c_{11}, c_{12}, c_{21}, and c_{22} are fixed values; and t_{max} is the maximum number of iterations.

Therefore, Eq. (5.14) should be changed as follows:

$$\begin{cases} V_m(t+1) = w(t)V_m(t) + c_1(t)r(t)(P_m(t) - X_m(t)) + c_2(t)r'(t)(G_d(t) - X_m(t)) \\ X_m(t+1) = X_m(t) + V_m(t+1) \end{cases} \tag{5.20}$$

When the position parameter of the particle is updated, it is necessary to calculate the fitness value of the new position and update the individual and population optimal position. The formulas are as follows:

$$P(t+1) = \begin{cases} X(t+1), \ F(X(t+1)) < F(P(t)) \\ P(t), \end{cases} \tag{5.21}$$

$$G(t+1) = \begin{cases} P_n(t+1), \ F(P_n(t+1)) < F(G(t)) \\ G(t), \end{cases} \tag{5.22}$$

where n represents the particle of best fitness. When the number of iterations reaches the maximum, the population optimal position G is the optimal solution.

(2) Results and analysis

In this paper, PSO algorithm is used to improve the BP neural network. After the weights and thresholds of BP neural network are optimized by the PSO algorithm, a model for the relationship between tag coordinates and reading distance has been set up by this improved neural network. There are 300 groups of data used for the experiment, each group has 7 tags. The results are shown in Table 5.1. (x_i, y_i, z_i) represents the spatial position of the i_{th} tag, d_r represents the actual reading distance,

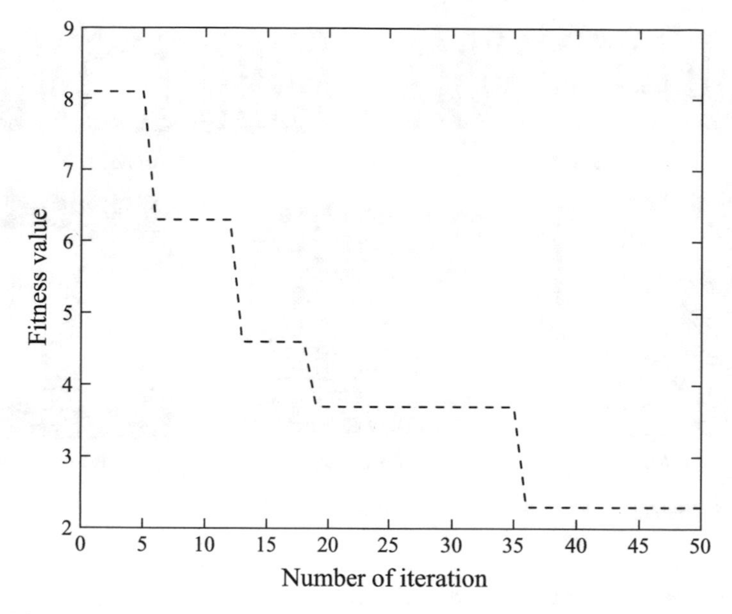

Fig. 5.8 The change of fitness value during the training

and d_p represents the predicted distance calculated by the model. *E* is the error, the formula is as follow

$$E = \frac{|d_p - d_r|}{d_r} \times 100\% \tag{5.23}$$

In this paper, the fitness value can also be used to evaluate the performance of the PSO neural network. The smaller the fitness value is, the better the performance of PSO neural network is Fig. 5.8 shows the change of fitness value during the training of PSO neural network. In the iterative process, the fitness value decreased from 8.1, and finally stabilized at 2.3. As mentioned above, the fitness value is the sum of errors, therefore, the average prediction error for each set of tags is less than 0.01.

From Table 5.1 and Figure 5.8, we can see that the model established by PSO neural network can simulate the relationship function between tag coordinate and reading distance well. Therefore, we can use this model to obtain the spatial distribution of the tags under the given distance.

We tested the optimization method with 5 different kinds of tag groups, the tags are shown in Fig. 5.9 Each kind of group has 300 sets of data obtained from the experimental measurements. The results show that the type of tag has no effect on this method. Therefore, the optimization method proposed in this paper can be applied to any type of tags.

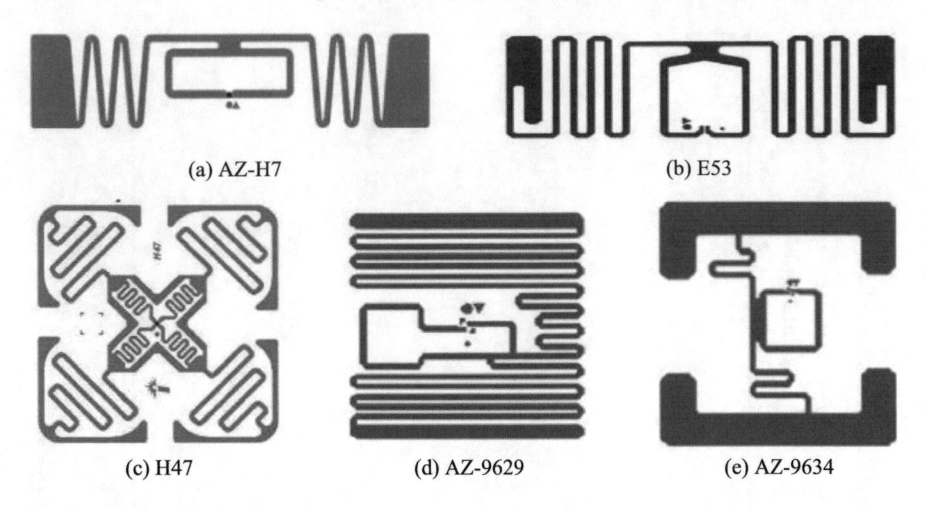

Fig. 5.9 5 different kinds of tags

(3) Comparison between PSO and GA-BP

In order to study the prediction performance of PSO and GA-BP neural network, we compared the prediction error of these two neural networks. The result is shown in Fig. 5.10. The slopes of the fitted linear function are 1.1(PSO) and 0.97(GA-BP). The curve with "*" represents the prediction error of PSO neural network, and the average error is 1.02%. The curve with "o" represents GA-BP, and the average error is 2.82%. Therefore, the comparison result shows that PSO neural network is more accurate than GA-BP neural network in prediction.

In addition, we also compared the uptime of PSO and GA-BP neural networks, and each neural network has been run for 200 times. The result is shown in Fig. 5.11. The

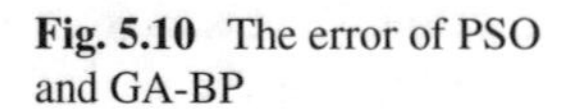

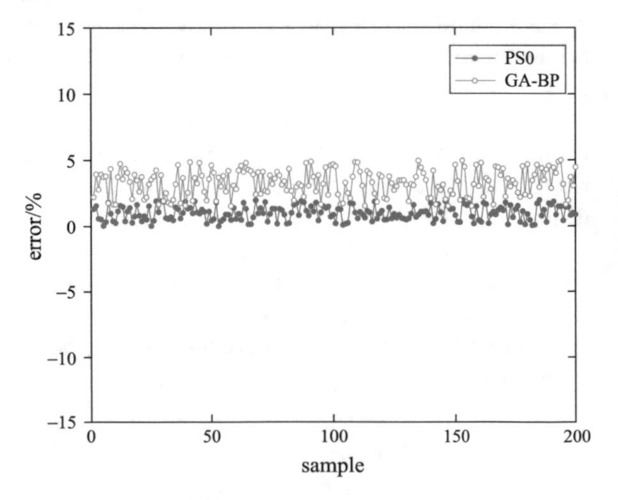

Fig. 5.10 The error of PSO and GA-BP

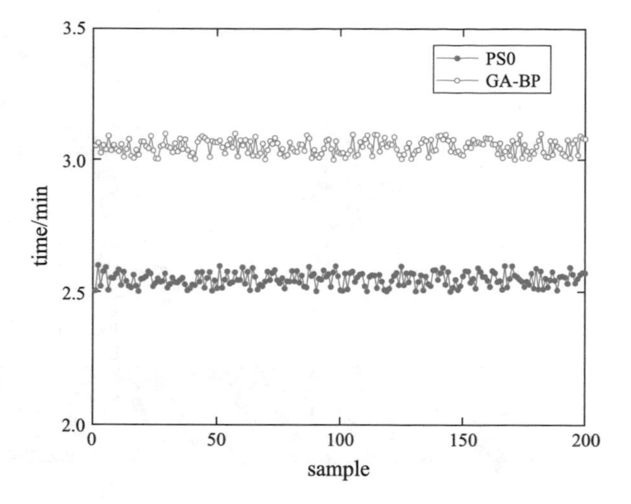

Fig. 5.11 The uptime of PSO and GA-BP neural network

average time GA-BP neural network spends on computation is 3.05min. However, PSO neural network only spends 2.55min. This means that PSO is not only better than GA-BP in prediction, but also in calculation.

5.2 Physical Anti-Collision Based on Support Vector Machine (SVM)

5.2.1 RFID Detection System

(1) Hardware constitution

In intelligent supply chain and asset management, RFID tags can hold many kinds of information about the products they are attached to, including serial numbers, configuration instructions, and much more. When the products arrive at unloading area, RFID readers installed in doors examine their contents and update the inventories of supply chain and asset management accordingly. Inside a warehouse, the products could be identified and tracked automatically. Once a product leaves the warehouse, the RFID readers check the contents of the tag attached to the product and update the inventories immediately.

For simulating the environment of products moving in and out, we design a RFID detection system, as shown in Figs. 5.12 and 5.13. The RFID detection system is mainly composed of a reader, reader antennas, an antenna stand, some tags, a tray, a laser ranging sensor, a transportation device, a charge coupled device (CCD), and a control computer. The application items of the system include the test of tags' reading range, anti-collision performance, and location optimization.

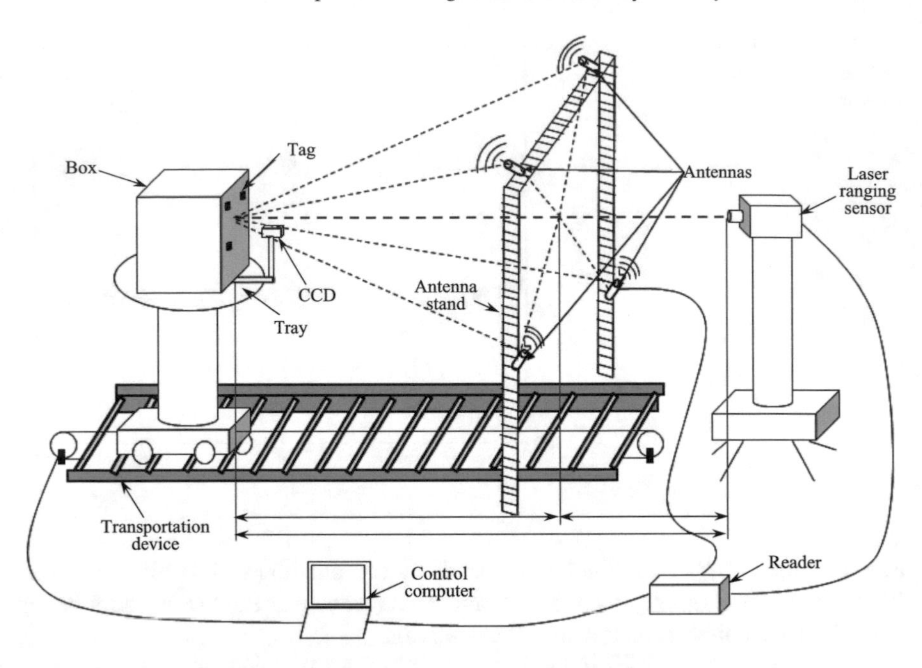

Fig. 5.12 Schematic diagram of RFID detection system (Color online)

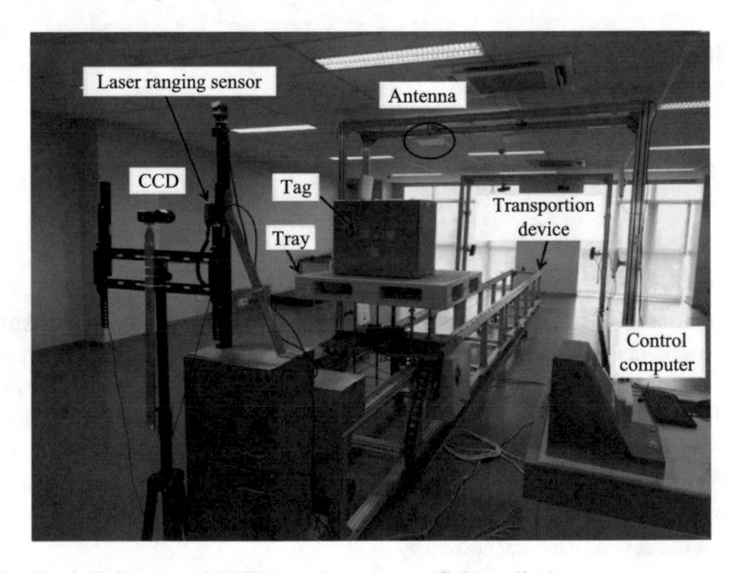

Fig. 5.13 Physical diagram of RFID detection system (Color online)

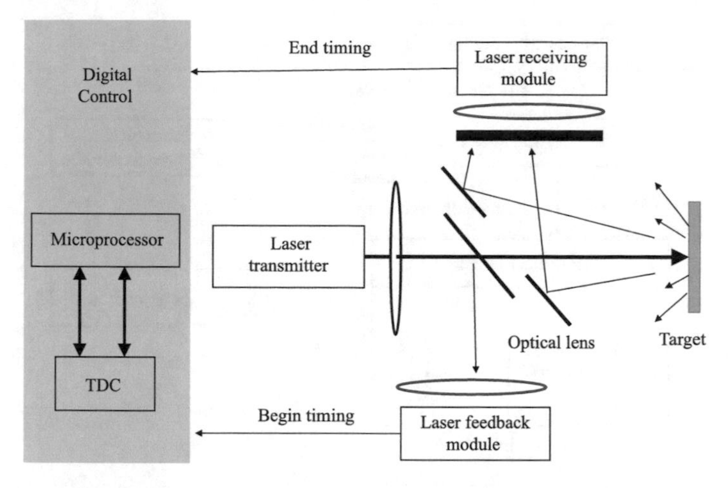

Fig. 5.14 Schematic diagram of laser ranging sensor

The tags are the commercial UHF tags Inlay H47 with its chip is Impinj Monza 4QT, the readers are Impinj speedway revolution R420, and the readers' antennas are Larid A9028. The laser ranging sensor is Wenglor X1TA101MHT88 with its measuring range is 30m. Simultaneously, the transmission coefficient of electromagnetic wave in a material has great relationship with the dielectric constant of material and the larger the dielectric constant, the lower the transmission coefficient. Therefore, we have chosen the carton as pasted material with low dielectric constant.

Due to the advantages of high accuracy, high speed, and good directivity, laser sensor has been widely applied in the field of distance measurement. In practice, there are two kinds of laser ranging sensor: pulse laser ranging sensor and phase laser ranging sensor. The pulse laser ranging sensor is chosen in this paper and the corresponding structure is shown in Fig. 5.14.

Firstly, laser transmitter sends a light pulse beam to measured target. When the optical pulse transfers in optical lens, it is divided into two parts. One beam, after reflected in the front of lens, transfers into the laser feedback module; subsequently, the beam is photovoltaic converted, filtered, rectified, and then transferred into time to digital converter (TDC). The other beam is diffused once it reaches the target through the lens; the part of diffuse light transfers into laser receiving module and is handled according to the above process. Finally, the whole measurement process is completed via processing the time difference of two beams.

In the whole system, the laser ranging sensor connects with RFID reader and control computer via serial port.

The system is applied to multiple readers and multiple tags. In the case of a reader and multiple tags, the reading performance could be improved via locating the tags appropriately. Simultaneously, in the case of multiple readers and random tags, the reading performance could be improved via locating the readers appropriately.

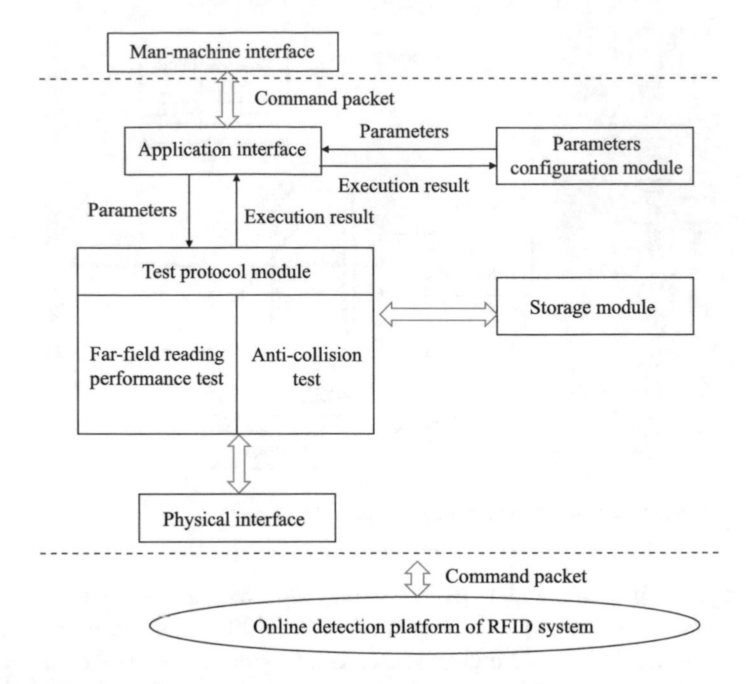

Fig. 5.15 Software architecture of detection system

(2) Software architecture
The software architecture of detection system consists of application interface module, parameters configuration module, test protocol module, and storage module, as shown in Fig. 5.15.

(1) Application interface module.
Application interface module plays a bridge role between test protocol module and man-machine interface. The deictic information of man-machine interface is transmitted to test protocol module by application interface module. The deictic information is integrated into the command packet according to which test protocol module and parameters configuration module complete the corresponding task.
(2) Parameters configuration module.
In initialization process, the parameters of RFID reader and antennas (IP address, read mode, search mode, output port, output level, and so on) are configured via parameters configuration module. In test process, some parameters (reader's transmitting power, antenna's receiving sensitivity, test times, and so on) also need to be configured.
(3) Test protocol module.
Test protocol module is the core of online detection software, and its main task is to complete various kinds of detection. At present, the main work is to test far-field reading performance and anti-collision of RFID system.

The test protocol module could be simultaneously extended according to test requirement. The test protocol module exchanges command data with other modules to realize parameter configuration, data storage and result shows.

(4) Storage module.

Once detection is over, the detection results are real-time shown and the data is stored in storage module.

(3) Theoretical analysis of test

In practical application, tags are attached to the surface of targets and generate induction current to send data due to the electromagnetic field transmitted by reader antennas. Then, the readers detect and decode the backscattered signal of tags. Eventually, the readers send the data of tags to the background processor, and the RFID system achieves the purpose of automatic identification of goods.

The power density of an electromagnetic wave incident on the RFID tag antenna in free space is given by

$$S = \frac{P_{tx}G_{tx}}{4\pi R^2} = \frac{P_{\mathrm{EIR}}}{4\pi R^2} \tag{5.24}$$

where P_{tx} is the transmitted power, G_{tx} is the gain of the reader's transmitting antenna, R is the distance to the tag, and P_{EIR} is the effective radiated power of transmitting antenna [18–20].

The power P_{tag}, collected by the tag antenna, is by definition the maximum power that can be delivered to the complex conjugate matched load:

$$P_{\mathrm{tag}} = A_e S = \frac{\lambda^2}{4\pi} G_{\mathrm{tag}} S = P_{tx} G_{tx} G_{\mathrm{tag}} \left(\frac{\lambda}{4\pi R}\right)^2 \tag{5.25}$$

where G_{tag} is the tag antenna's gain, A_e is the effective area of the tag's antenna given by

$$A_e = \frac{\lambda^2}{4\pi} G_{\mathrm{tag}} \tag{5.26}$$

The power backscattered from the tag is expressed by

$$P_{\mathrm{back}} = S\sigma = \frac{P_{tx}G_{tx}}{4\pi R^2}\sigma = \frac{P_{\mathrm{EIR}}}{4\pi R^2}\sigma \tag{5.27}$$

The power received by the receiving antenna of reader can be calculated from the classical radar equation as:

$$P_{rx} = A_W S_{\text{back}} = \frac{P_{tx} G_{tx} G_{rx} \lambda^2}{(4\pi)^3 R^4} \sigma \tag{5.28}$$

where σ is the radar cross section of the RFID tag, G_{tx} is the gain of the reader's receiving antenna, and A_W is the effective area of the reader's antenna given by

$$A_W = \frac{\lambda^2}{4\pi} G_{rx} \tag{5.29}$$

(4) Test process

The following is the procedure of detection, as shown in Fig. 5.16. Initially, a tray is installed on a transportation device and some boxes with tags are placed on the tray. Moreover, the distribution image of tags is collected via CCD. Then, a RFID reader and a plurality of RFID antenna are installed on the antenna stand, and the beam of laser ranging sensor points to the boxes. Afterwards, the cycle index of tray is set, and the tray transports boxes on the transportation device at a certain speed to simulate the goods in-out warehouse. When the tagged boxes enter the read zone of the reader antennas, the antennas receive the RF signal of tags and the reader

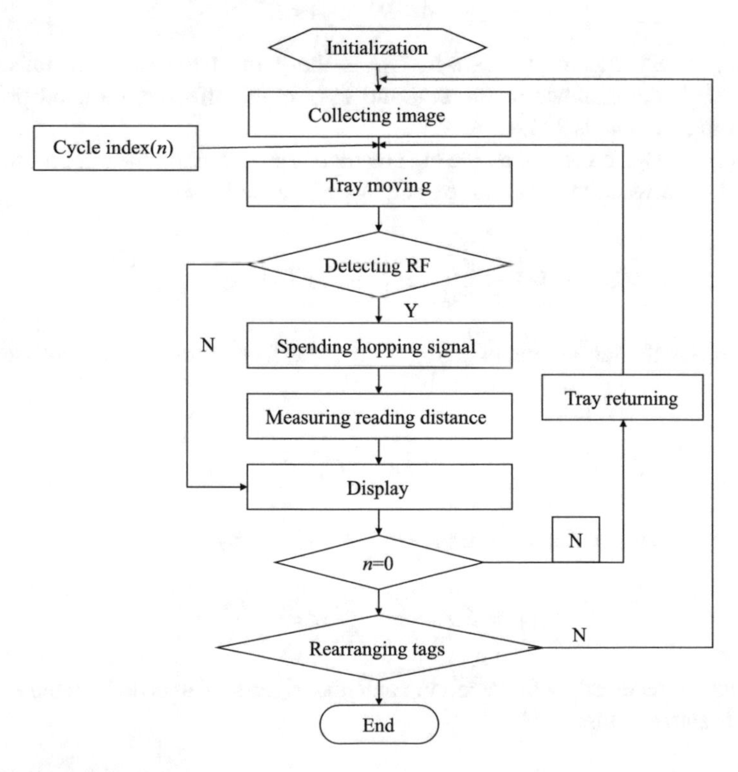

Fig. 5.16 Procedure of detection

sends a hopping signal to activate the laser ranging sensor. Consequently, we could calculate the average reading distance of RFID antenna to tags when the cycle index reaches the set value. Eventually, the tray returns to initial point and repeats the above operation in the case of changing the distribution of tags.

The measurement of reading distance is indirect to the survey. Adjusting the optical lifting platform to ensure the laser beam of laser ranging sensor aiming at the boxes, we define the intersections of laser ranging sensor's beam and antenna stand plane to reference point.

The distance of tags to reference point is

$$T = S - L \tag{5.30}$$

where L is the deterministic distance of laser ranging sensor to reference point and S is the distance of laser ranging sensor to tags.

The distance of ith RFID antenna to tag is

$$R_i = \left(T^2 + H_i^2\right)^{1/2} \tag{5.31}$$

where H_i is the distance of ith RFID antenna to reference point. Once the antennas of reader are fixed on the antenna stand, the distance of ith RFID antenna to reference point is measured manually and then input to the main program. In whole measurement, the distance is constant. In practice, the boxes are set to the same and the tags are on the front of the boxes.

5.2.2 *Position Location of Tags*

The purpose of image recognition is to identify target automatically. In image recognition system, the quality of images is improved and the target is identified or classified. The mathematical morphology is used to extract image components that are useful in the representation and description of region shape (such as boundaries, skeletons, and convex hull).

Dilation and erosion are the basic elements of mathematical morphology. In this paper, we use dilation and erosion to remove the noise of image. With A and B as sets in Z^2, the dilation of A by B, denoted $A \oplus B$, is defined as [30]

$$A \oplus B = \{z|(B)_z \cap A \neq \emptyset\} \tag{5.32}$$

For sets A and B in Z^2, the erosion of $A \Theta B$ is defined as

$$A \Theta B = \{z|(B)_z \subseteq A\} \tag{5.33}$$

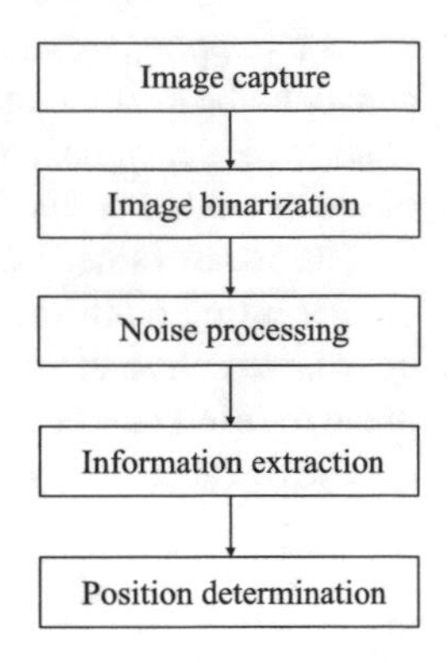

Fig. 5.17 Procedure of tags' position locating

In the part of feature extraction and recognition, we get the location of tags based on shape feature, as shown in Fig. 5.17.

Initially, the collected image is preprocessed to convert to binary image and then removed background noise. Afterwards, the information (area, perimeter, and so on) of the image is obtained. Eventually, the position of coordinate origin and tags are derived based on shape index. In this paper, three tags are chosen to test, as shown in Figure 5.18.

For the purposes of image matching, the binary image needs to be described based on shape index. The shape index refers to the ratio of tag's SQ (squares of perimeter) and its area:

$$C = \frac{P^2}{A} \tag{5.34}$$

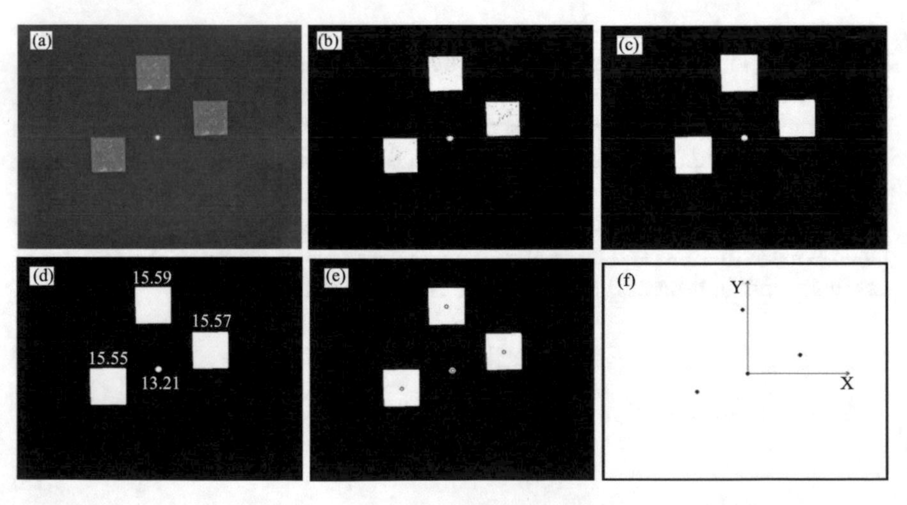

Fig. 5.18 Position locating of tags. **a** is collected image, **b** is binary image, **c** is the image removed noise, **d** is the shape index of tags, **e** is center mark, **f** is the position of coordinate origin and tags

where P is the perimeter of tag image and A is its area. The more the value of C, the more complicated the shape of tag. Therefore, the shape index of RFID tag is roughly 16 and the shape index of laser point is roughly 12.56, as shown in Fig. 5.18d. According to the marked tags and laser point, the position of coordinate and tags could be obtained, as shown in Fig. 5.18e, d.

5.2.3 Predict Model of RFID Tags' Distribution Based on SVM

The dynamic reading performance of RFID multiple tags system is not only influenced by algorithm, but also affected by tags' geometry distribution. Establishing geometric model and predicting optimal geometric distribution using SVM could improve reading performance of RFID system.

(1) SVM regression algorithm

The support vector machine is established at the foundations of Vapnik-Chervonenkis (VC) dimension and structural risk minimization (SRM) [31].

Consider one has a training data set $\{(x_i, y_i)|i = 1, 2, \ldots k\}$, where x_i are the training examples and y_i are the class labels. The class label of x is obtained by considering the sign of $y - f(x)$. Consider using $g(x) = \omega \cdot x + b$ to fit the training examples realizing that the distance of f and g is minimum. In other words, the loss function $R(f, g) = \int L(f, g)\mathrm{d}x$ is minimum. According to structural risk minimization,

$$J = \frac{1}{2}\|\omega\|^2 + C\sum_{i=1}^{k} L(g(x_i), y_i) \tag{5.35}$$

The optimization problem can be written as

$$\min\left[\frac{1}{2}\|\omega\|^2 + C\sum_{i=1}^{k}(\xi_i, \xi_i^*)\right]$$
$$s.t.\begin{cases} y_i - \omega \cdot x - b \le \varepsilon + \xi_i \\ \omega \cdot x + b + y_i \le \varepsilon + \xi_i^* \\ \xi_i, \xi_i^* \ge 0 \end{cases} \tag{5.36}$$

where $\varepsilon > 0$ is fitting precision, ξ_i is the value above target, ξ_i^* is the value under target, and $C > 0$ is a constant. The solution of this problem is obtained using the Lagrange theory:

$$\max\left[-\frac{1}{2}\sum_{i,j=1}^{k}(\alpha_i-\alpha_i^*)(\alpha_j-\alpha_j^*)(x_i\cdot x_j)\right.$$
$$\left.-\varepsilon\sum_{i=1}^{k}(\alpha_i+\alpha_i^*)+\sum_{i=1}^{k}y_i(\alpha_i^*-\alpha_i)\right]$$
$$s.t.\begin{cases}\sum\limits_{i,j=1}^{k}(\alpha_i-\alpha_i^*)=0\\ \alpha_i,\alpha_i^*\in[0,C]\end{cases} \tag{5.37}$$

where α_i and α_i^* are Lagrange factors.

Regression function could be changed into high dimension via the kernel function $K(x_i\cdot x_j)$:

$$f(x)=\omega\cdot x+b=\sum_{i,j=1}^{k}(\alpha_i^*-\alpha_i)K(x_i\cdot x_j)+b^* \tag{5.38}$$

(2) Predicted reading distance of tags based on SVM

In this paper, we analyzed the position of tags and the corresponding reading distances. There are 200 groups of samples and each sample has 6 characteristics, as shown in Table 5.2. The data is taken as training samples and test samples simultaneously.

In Table 5.2, x and y represent the horizontal and vertical coordinates of tags, respectively. The training data of 200 groups are normalized and then the cross-validation method is used to get the optimal parameters $c=4$, $g=0.125$ of SVM, in which c is the penalty coefficient and g is the kernel function coefficient [32]. The parameter selection results are shown in Fig. 5.19.

We use the optimal parameters (c and g) to establish regression model and predict results. Finally, the results are anti-normalized, as showed in last two columns of

Table 5.2 Test sample data

x_1/m	y_1/m	x_2/m	y_2/m	x_3/m	y_3/m	R_m/m	R_p/m	η/ %
0.031	0.105	0.132	0.043	−0.092	−0.034	2.14	2.15	−0.47
0.051	0.176	0.032	0.145	−0.107	0.125	2.37	2.34	1.26
−0.065	0.048	0.149	0.128	−0.139	−0.086	2.45	2.47	−0.82
⋮	⋮	⋮	⋮	⋮	⋮	⋮	⋮	⋮
0.128	0.036	0.067	−0.107	−0.089	0.096	1.97	1.96	0.51
−0.078	0.063	0.032	−0.139	−0.128	−0.137	2.62	2.61	0.38
0.021	0.076	0.112	0.087	0.012	0.035	2.48	2.49	−0.40

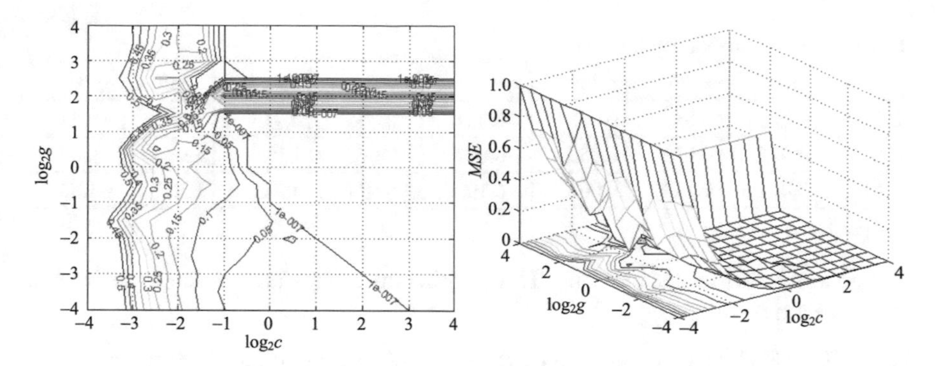

Fig. 5.19 Subtle parameter selection

Table 5.2. R_m is true reading distance, R_p is predicted reading distance, and the relative prediction error is given by

$$\eta = \frac{R_m - R_p}{R_m} \quad (5.39)$$

Simultaneously, the corresponding relative prediction error with the change of reading distance is shown in Fig. 5.20, in which MSE is 4.79×10^{-5} and correlation coefficient $R = 99.9629\%$.

(3) Experimental verification of optimal geometric distribution of RFID tags

Through analyzing the predictive value, we could obtain the maximum and minimum reading distance and then verify the experimental results. The value of reading distance is maximum (2.92 m) at 2.5×10^4 groups and minimum (1.53 m) at $4.8 \times$

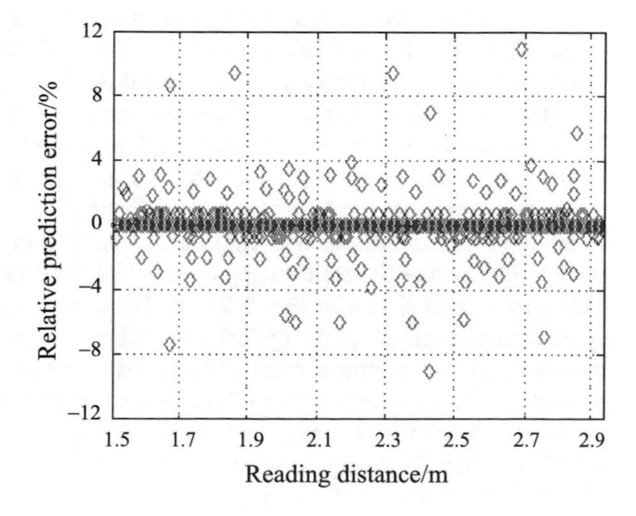

Fig. 5.20 Relative error of predicted results

Table 5.3 Experimental verification

x_1/m	y_1/m	x_2/m	y_2/m	x_3/m	y_3/m	R_m/m	R_p/m	η/ %
0.010	0.102	0.083	0.053	−0.087	−0.055	2.91	2.92	−0.34
0.008	0.095	0.083	0.052	−0.092	−0.063	2.90	2.92	−0.68
0.012	0.004	0.112	0.002	0.102	0.011	1.51	1.53	−1.32

10^3 groups in 1.7×10^{11} groups. Three particular groups are chosen to verify and test steps are as follows:

(1) The RFID tags are attached on the goods according to test distribution.
(2) The cycle indexes of tray are set 10 times and the experimental data is statistical average to ensure the reliability of reading distance.
(3) The reading distances of different tags' location are measured.

Experimental results are shown in Table 5.3. Experimental results show that the optimal geometry distribution is not only regular polygon, but also non-simple polygon. Simultaneously, the predict values equate with the measurement results. The method of using SVM to predict the optimal geometric distribution of RFID tags is feasible.

5.3 Physical Anti-Collision Based on Wavelet

5.3.1 Design of Image Analysis System

(1) Construction of measurement platform

Figure 5.21 shows the schematic diagram of the dual CCD semi-physical verification test platform, and Fig. 5.22 shows its picture. Dual CCD semi-physical verification test platform mainly consists of reader, antenna, RFID tag, RFID tag bracket, control-computer, servo motor, vertical camera, horizontal camera, guide rail, and turntable. RFID tag brackets are labeled at the bottom. The reader is connected with the antenna and the control-computer respectively. Vertical camera and horizontal camera are also connected with the control computer, respectively.

Semi-physical verification test platform consists of image acquisition system and tag reading distance dynamic testing system. The RFID tag reading distance dynamic testing system is shown in Fig. 5.23. RFID tag reading distance dynamic testing system consists of two parts: optical lifting platform and laser range finder. Firstly, the optical lifting platform is adjusted so that the laser beam is aligned with the geometric center of the tag group. Then, the laser ranging sensor is used to test the distance between the tag group and the antenna when every tag of the tag group is identified. UHF RFID tags-H47 are used as RFID tags. CCD camera has 1288 × 964 pixels and the size of each pixel is 3.75 × 3.75 μm. The CCD camera's frame

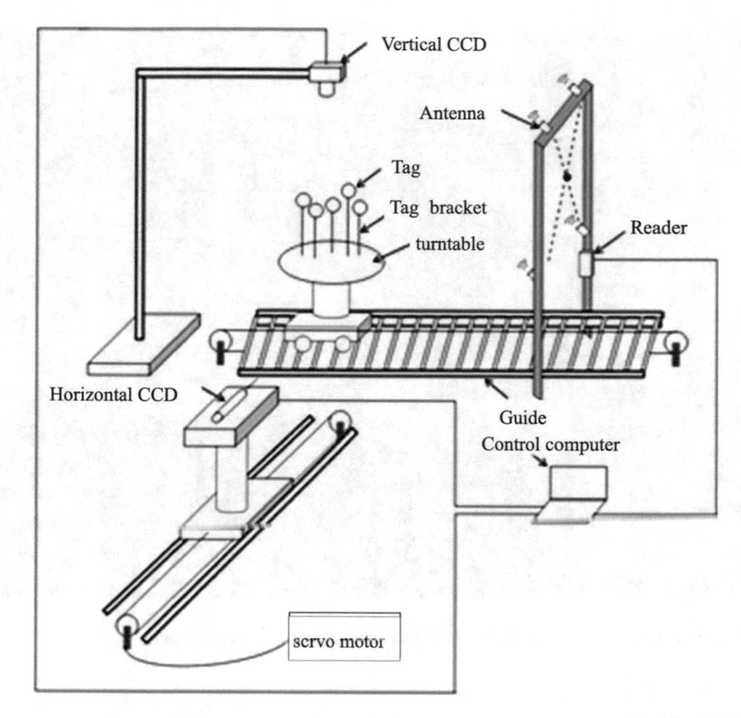

Fig. 5.21 Dual CCD tag image acquisition system schematic

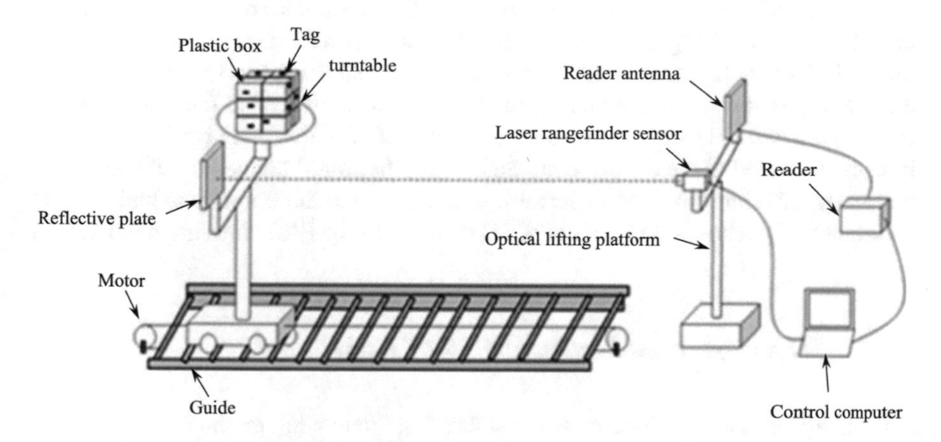

Fig. 5.22 Schematic diagram of RFID tag reading distance dynamic testing system

rate is 30 FPS. Impinj Speedway Revolution R420 reader and Laird A9028 far-field antenna are used as the reader and reader antenna in the system. The maximum RF output power of the reader antenna is 30dBm.

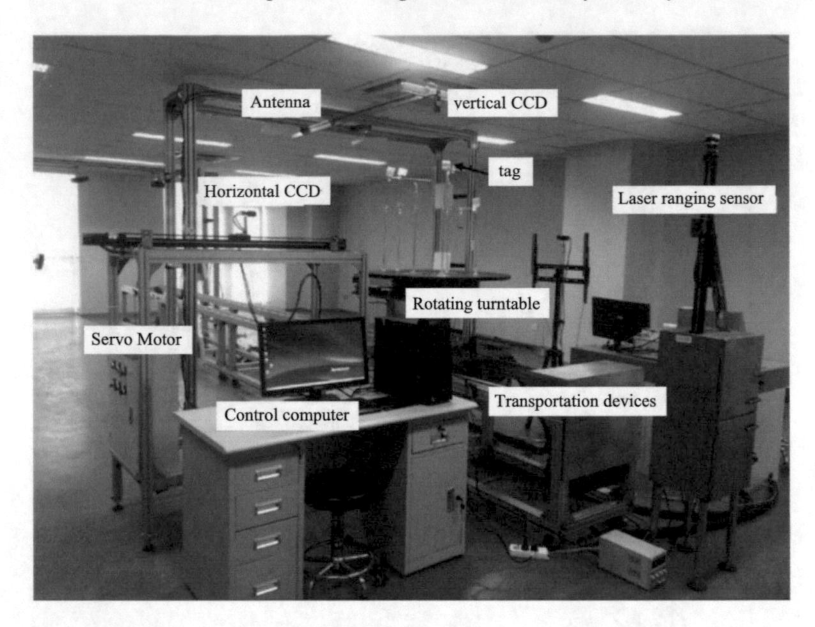

Fig. 5.23 Practicality picture of semi-physical verification test platform

(2) Test process

The whole test platform is used to measure the 3D coordinates of RFID tags, and the test flow is shown in Fig. 5.24. A number of brackets with RFID tags are placed on the turntable. The RFID tag reading distance dynamic testing system is started to obtain the reading distance corresponding to the tag distribution. The control computer controls the vertical camera to obtain the images of turntable and tags. The obtained images are stored in the control computer. Then, the stored images are denoised and the image matching process is performed to obtain the 2D coordinate and vertical coordinate of each RFID tag. Finally, 3D coordinates of RFID tags are obtained.

5.3.2 3D Coordinate Measurement of RFID Tag

(1) Image denoising based on wavelet threshold denoising method

In the process of digital image acquisition and transmission, there are varieties of noise in the image caused by many factors such as light intensity and temperature of sensor, which seriously affects the processing and analysis of the received images. In order to make better use of the images to obtain 3D coordinates of the RFID tags, it is necessary to denoise the received images.

Wavelet analysis has the characteristics of multi-resolution analysis and the ability to characterize the signal in the time domain and frequency domain. Wavelet analysis

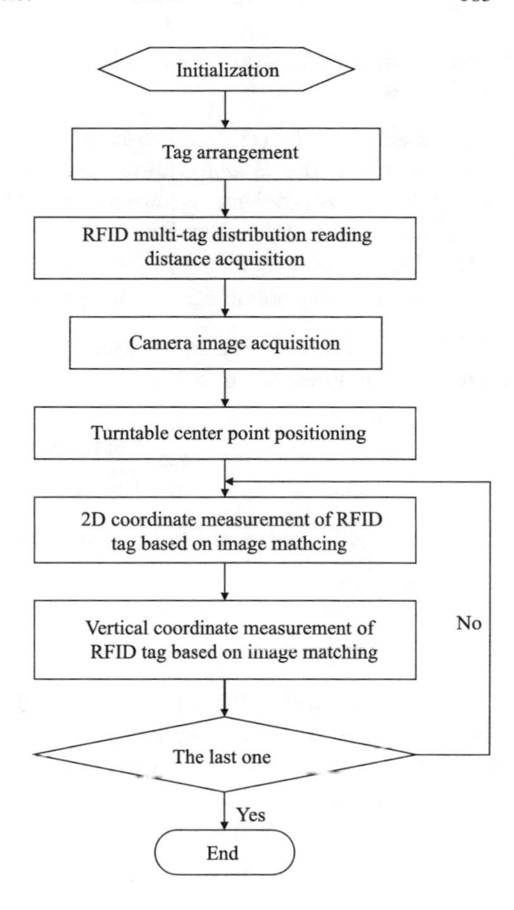

Fig. 5.24 Test flow chart

has become a common frequency domain analysis tool for signal processing. The principle of wavelet threshold denoising method is based on different properties of wavelet coefficients in different scales between signal and noise. By using mathematical tool to select the corresponding coefficients, the wavelet threshold denoising method can process the wavelet coefficients of noisy signals. The basic process is shown in Fig. 5.25.

The basic idea of wavelet threshold denoising method is to set a critical threshold λ. If wavelet coefficients are less than λ, it is considered that the coefficients are mainly caused by noise, and the coefficients will be removed. If the wavelet coefficients are larger than λ, the coefficients are considered to be mainly caused by the signal, and the coefficients will be obtained. Finally, the wavelet coefficients are transformed by

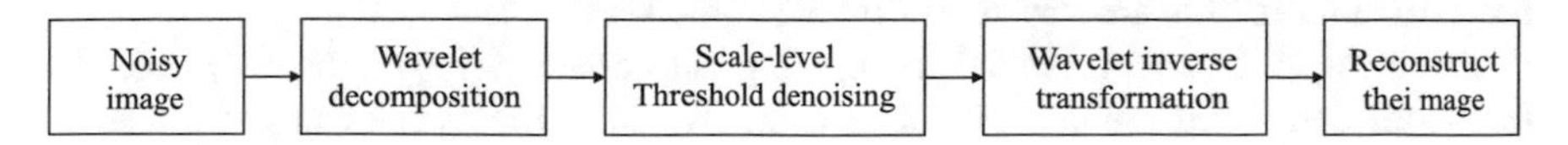

Fig. 5.25 Wavelet denoising flow chart

wavelet inverse transformation to obtain the denoised signals. The detailed steps are as follows:

(1) Noisy signal $f(t)$ is decomposed by wavelet transform method. Furthermore, we obtain the decomposition coefficients $w_{j,k}$.
(2) Wavelet coefficients $w_{j,k}$ are processed by threshold λ. The processed wavelet coefficients are recorded as $\bar{w}_{j,k}$.
(3) The wavelet coefficients $\overline{w}_{j,k}$ are reconstructed by the wavelet inverse transform to obtain estimation signal $\overline{f}(t)$, which is the denoised signal.

The soft threshold method is used to select the wavelet transform coefficients. The equation is shown in (5.40).

$$f(x) = \begin{cases} \operatorname{sgn}(x)(|x| - \lambda) & |x| > \lambda \\ 0 & |x| \le \lambda \end{cases} \tag{5.40}$$

In Eq. (5.40), sgn(x) stands for a symbolic function which is shown in Eq. (5.41).

$$\operatorname{sgn}(x) = \begin{cases} 1 & x > 0 \\ -1 & x < 0 \end{cases} \tag{5.41}$$

In order to verify the effectiveness and superiority of the wavelet threshold denoising algorithm, a noisy image obtained in a typical scene is selected, and different methods are used to denoise it, respectively. Experimental results are shown in Fig. 5.26 and Table 5.4.

The equation of SNR is shown in (5.42).

$$\mathrm{SNR} = 10\lg\left[\frac{\sum_{i=1}^{M}\sum_{j=1}^{N} g(i,j)^2}{\sum_{i=1}^{M}\sum_{j=1}^{N} [g(i,j) - f(i,j)]^2}\right] \tag{5.42}$$

In Eq. (5.42), M represents the image's length. N represents the image's width. $f(i, j)$ and $g(i, j)$, respectively, represent the gray values of the original image and the denoised image.

Figure 5.27 and Table 5.5 show that the denoising method based on sym4 wavelet threshold is superior to other denoising methods, either from subjective visual effects or from objective quality evaluation criteria. In order to further verify the validity and practicability of this algorithm, different images of different scenes and the same noise degree of different images are used to conduct denoising experiments. Experimental results are shown in Fig. 5.27 and Table 5.5.

From Fig. 5.27 and Table 5.5, comparing with other methods, the sym4 wavelet threshold denoising method performs best. By using the sym4 wavelet threshold denoising method, the quality of the noisy images has been greatly improved not only from the subjective visual aspect but also from the objective quality evaluation aspect.

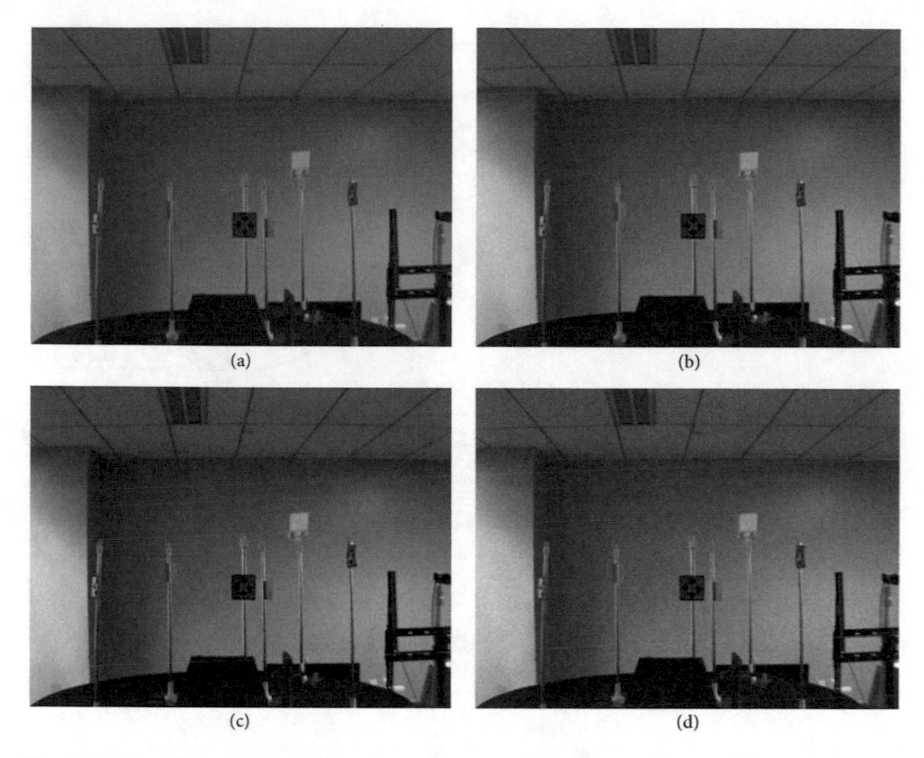

Fig. 5.26 Denoising results of different denoising methods. **a** Wavelet threshold denoising results by using db2 function. **b** Wavelet threshold denoising results by using sym4 function. **c** Wiener filter denoising results. **d** Median filter denoising results

Table 5.4 Objective quality evaluation of denoising methods results

Denoising methods	SNR/dB
Experimental initial image	5.3862
Wavelet threshold denoising method by using db2 function	13.3086
Wavelet threshold denoising method by using sym4 function	13.3132
Wiener filter denoising method	13.1488
Median filter denoising method	11.3075

Therefore, the author uses sym4 wavelet threshold denoising method to denoise the noisy images.

(2) Image matching

After denoising noisy images, in order to conduct the 3D coordinates measurement of the RFID tags, the obtained images of the RFID tags are matched. In this paper, a template matching method is used to match the tag images. Firstly, we select an

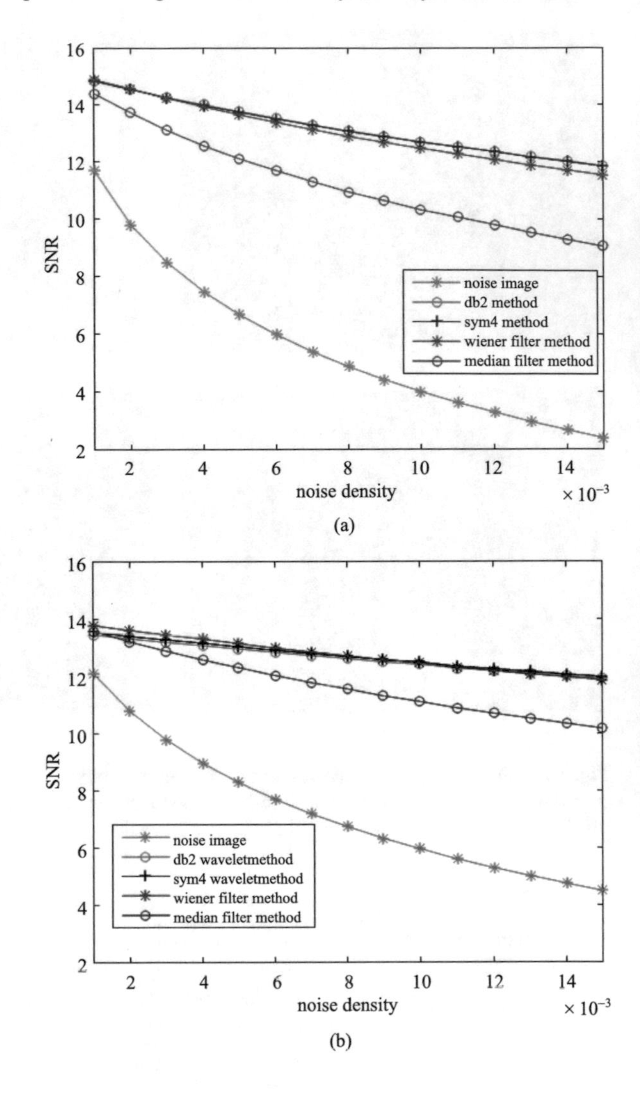

Fig. 5.27 Denoising effects of different scenes. **a** scene 1. **b** scene 2

image as the matching template (assuming its size is $m \times n$ pixels). Then, we select a target point in the target image. With the target point as the center, $m \times n$ pixels gray array is selected as the target matching area. After that, all $m \times n$ pixel gray-scale arrays in the target image are taken out sequentially. The correlation coefficient between the matching area and the template is calculated one by one. The similarity of matching area is determined by the correlation coefficient. When the correlation coefficient is max, the matching area is the matching region.

As shown in Fig. 5.28, by template matching, a matching region can be found in the image.

Table 5.5 Denoising results of different scenes

	Original noisy image	db2 method	sym4 method	Wiener filter method	Median filter method
scene 1	7.7324	16.2199	16.2746	16.0556	13.9894
scene 2	6.5652	14.0288	14.0775	13.9889	12.3192
scene 3	6.0982	14.3203	14.3815	14.0722	12.4478
scene 4	5.3621	13.7378	13.7803	13.6026	11.7055
scene 5	6.5006	14.4023	14.4536	14.3226	12.4982
scene 6	5.7315	13.7388	13.8019	13.5380	11.9579
scene 7	6.1654	14.7818	14.8242	14.6252	12.4754
scene 8	6.6532	13.8800	13.9356	13.8672	12.2909
scene 9	5.2235	12.5797	12.6294	12.4445	11.0054
scene 10	6.0594	13.8803	13.9328	13.6897	11.9642

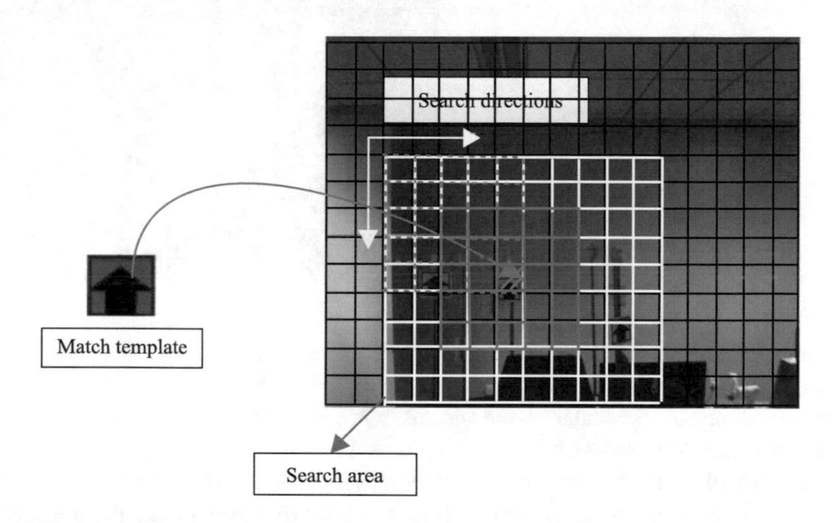

Fig. 5.28 Diagrams of template matching

(3) RFID tag 2D coordinate measurement

After denoising process, in order to obtain the 2D coordinate of RFID tag, the Canny edge detection operator is used to process the image. The Canny edge detection operator firstly extracts the edge profile of the turntable. Then two arcs are taken on the edge contour and vertical bisectors of arcs are calculated. The intersection of the vertical bisectors is the center of the turntable. As shown in Figure 5.29, with the center of the turntable as the origin point, the polar coordinate system is built with the direction of the horizontal camera as the polar axis.

The RFID tags are placed on the turntable and the mark point is pasted at the bottom of each tag. After that, one of the marks is selected as the template and each

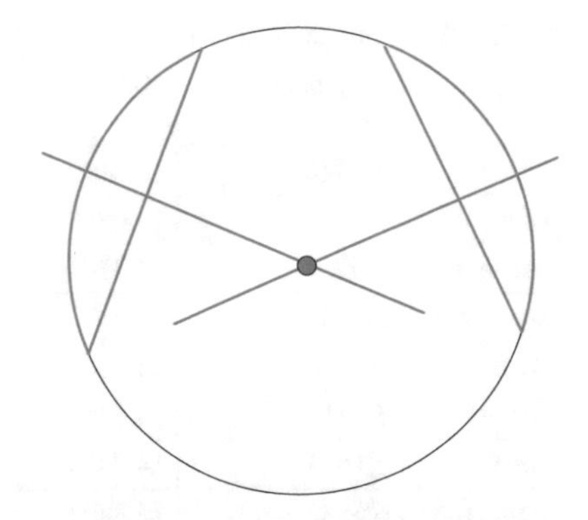

Fig. 5.29 Diagram of center location of turntable

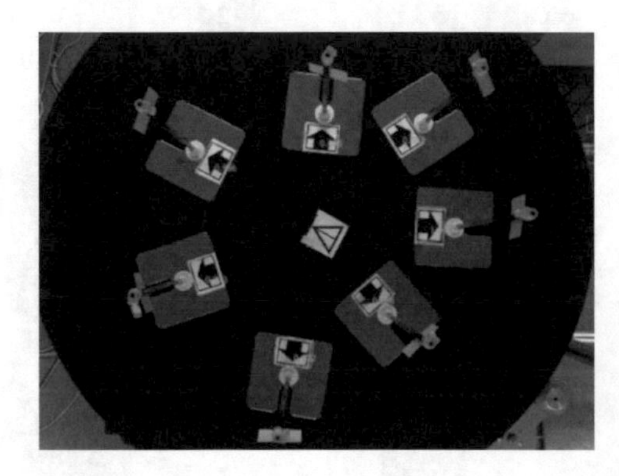

Fig. 5.30 Graph of match results

tag's mark is matched with the template. The matching result is shown in Fig. 5.30. By template matching, the position of each tag on the turntable is found. The distance from each mark to the center of the turntable is calculated and recorded as r_i. The subscript i represents the i_{th} tag.

The control computer controls the servomotor to drive the turntable to rotate. The control computer controls the horizontal camera to adjust so that the horizontal camera can focuse the i_{th} tag accurately. The i_{th} RFID tag's side view image is obtained, and the obtained image is stored in the control computer. The angle which the i_{th} tag rotates is the polar angle θ_i. The θ_i and r_i are the horizontal 2D coordinate parameter of the i_{th} RFID tag. The horizontal 2D coordinate of the i_{th} RFID tag is further calculated as $(r_i \cos\theta_i, r_i \sin\theta_i)$. Repeat the above steps in Sect. 3.3 to obtain every tag's 2D coordinate.

(4) RFID tag vertical coordinate measurement

(1) Controlling of horizontal camera

During the measurement of the vertical coordinate of the tag, the horizontal camera will automatically focus on the RFID tags. Firstly, the servomotor drives the turntable to rotate so that the horizontal camera can face the i_{th} RFID tag. Assuming that the horizontal distance between the horizontal camera and the center of the turntable is L, the horizontal distance d_i between the tag and the horizontal camera can be obtained in Eq. (5.43).

$$d_i = L - r_i \tag{5.43}$$

Then, the theoretical distance l_i which the horizontal camera needs so that the horizontal camera can exactly focus on the i_{th} RFID tag is calculated in Eq. (5.44).

$$l_i = \frac{fl'}{l' - f} \tag{5.44}$$

where l' is the distance between the lens and CCD sensor inside horizontal camera, f is the focal length of the horizontal camera.
Finally, the distance that the horizontal camera needs to move can be calculated in Eq. (5.45).

$$\Delta L_i = d_i - l_i \tag{5.45}$$

If ΔL_i is larger than zero, the horizontal camera approaches the tag. Otherwise, the horizontal camera moves away from the tag.

(2) RFID tag 2D coordinate measurement.

After obtaining the 2D coordinates of the tags, we choose one tag as the temple. Then, the control-computer controls the servomotor to drive the turntable to rotate. The control computer adjusts the horizontal camera to obtain the template's image. The vertical distance from the center of the template to the turntable is calculated as the vertical coordinate of the template. After taking the vertical coordinate of the template as the reference zero in vertical direction, the control computer controls the servomotor to drive the turntable to rotate. The control computer adjusts the horizontal camera to get every tag's image. Each tag is matched with the template by using the template matching method, and part of the tags matching results are shown in Fig. 5.31.
After successful matching, the vertical distance of the template is used as the reference zero to calculate the number of pixels between the center of the tag and the center of the template. Because each pixel in the CCD size is known and the horizontal distance from image plane to horizontal camera is a fixed value, the size of each pixel in the image can be calculated by using the triangular similarity principle. The vertical coordinate of the i_{th} RFID tag in the vertical direction is calculated by the tag's side view image and the difference number of pixels in the vertical direction between the i_{th} RFID tag and the template. The equation is:

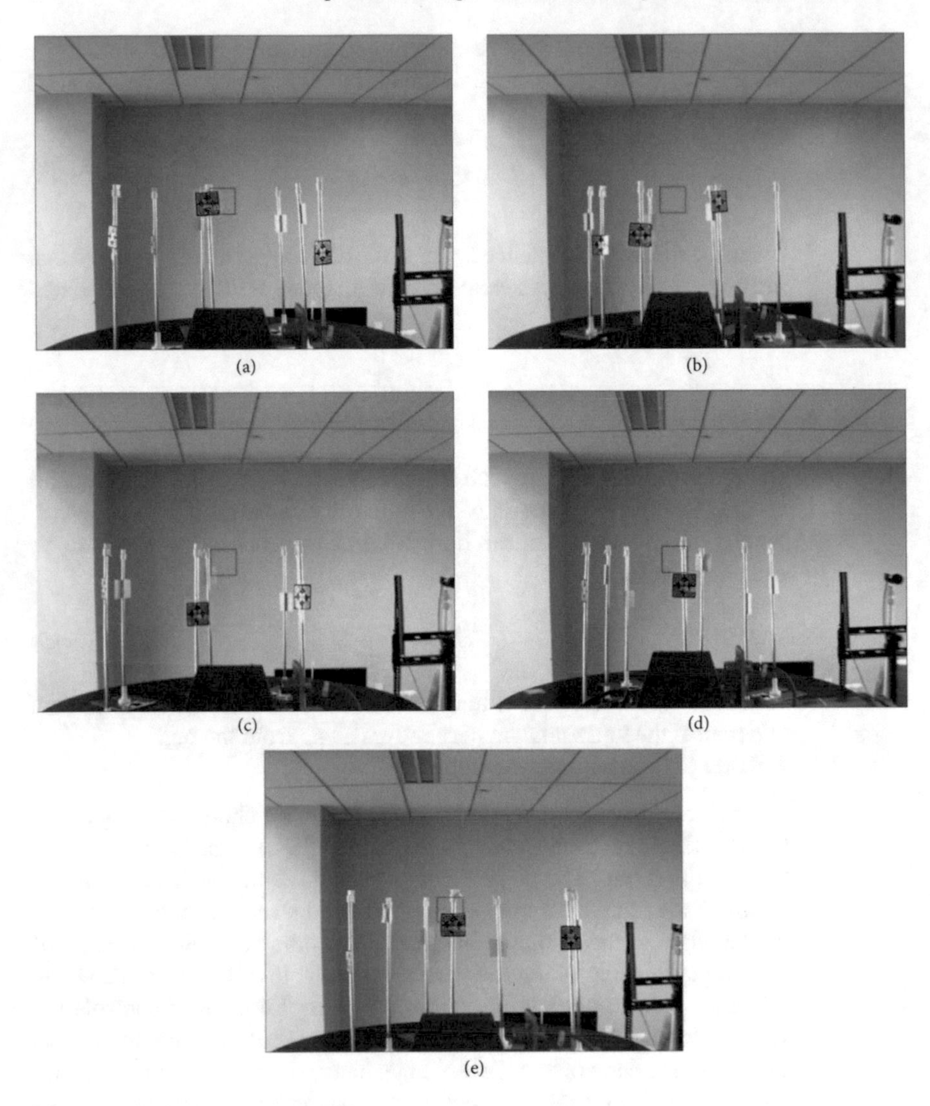

Fig. 5.31 Graphs of tags matching results

$$H_i = h \pm c_i \times a \tag{5.46}$$

In Eq. (5.46), H_i is the vertical coordinate of the i_{th} tag. a is the length of the image pixel.c_i is the difference number of pixels between the center point of the i_{th} RFID tag and the center point of the template. When the center point of the i_{th} RFID tag is below the center point of the template, Eq. (5.46) takes the "−" sign. When the center point of the i_{th} RFID tag is above the center point of the template, Eq. (5.46) takes the "+" sign.

Repeat the above steps in Section 5.3.4. Each detected tag is matched with the template, respectively, and the relative distance between the tag and the template in the vertical direction is calculated. Then, we obtain the vertical coordinate of each tag.

5.3.3 *RFID Multi-tag Wavelet Neural Network*

(1) Construction of wavelet neural network model

A large number of test data show that there is a complex nonlinear relationship between the 3D coordinate distribution of RFID multi-tag network and the corresponding reading distance. Some scholars have used GA-BP to model the multi-tag network [33]. In this paper, a new RFID multi-tag wavelet neural network model is proposed. The wavelet neural network model does not need to construct a mathematical function-based model, and it can approximate any nonlinear function. Therefore, the wavelet neural network is used to model the complex nonlinear relationship between the 3D coordinate distribution of RFID multi-tag network and the corresponding reading distance. The wavelet neural network is an improved neural network. The wavelet basis function is used as the transfer function of the hidden node. The error of the signal is transmitted back at the same time as the forward propagation.

Assuming $X_1, X_2, \ldots, X_k$ are the input parameters of wavelet neural networks. $Y_1, Y_2, \ldots, Y_m$ are predictive outputs of wavelet neural networks. Ω_{ij} and ω_{jk} are wavelet neural network weights. When the input signal sequence is $x_i(i = 1, 2, \ldots, k)$, the output of hidden layer is shown in Eq. (5.47).

$$h(j) = h_j\left[\frac{\sum_{i=1}^{k}\omega_{ij} - b_j}{a_j}\right] \quad j = 1, 2, 3, \cdots, l \tag{5.47}$$

In Eq. (5.47), $h(j)$ is output value for the j_{th} node of the hidden layer. ω_{ij} is the connection weight between the input layer and the hidden layer. H_j is the wavelet basis function. B_j is translation factor of wavelet basis function h_j. a_j is stretching factor of wavelet basis function h_j.

In this paper, we use the Morlet mother wavelet basis function as the activation function. The equation of wavelet basis function is shown in Eq. (5.48).

$$y = \cos(1.75x)e^{-x^2/2} \tag{5.48}$$

The output layer of wavelet neural network is calculated in Eq. (5.49).

$$y(k) = \sum_{i=1}^{l} \omega_{ik} h(i) \quad k = 1, 2, \ldots.., m \tag{5.49}$$

In Eq. (5.49), ω_{ik} is the weight from hidden layer to output layer. $h(i)$ is the i_{th} output of hidden layer nodes. l is the number of hidden layer nodes. m is the number of output layer nodes.

Wavelet neural network weight parameter correction algorithm is similar to BP neural network weight correction algorithm, using gradient correction method to modify the weight of network and the parameters of wavelet basis function so that the wavelet neural network prediction output is approaching the desired output.

(2) Model structure design

In order to improve the performance of the network, this paper selects 400 sets of RFID multi-tag 3D coordinate distribution data and the corresponding reading distance data as training samples. A three-layer wavelet neural network with an input layer, a hidden layer, and an output layer is used to approximate the relation model. At first, fewer hidden nodes are used to train the network. If requirements are not reached, the number of hidden nodes increases and the neural network is trained until requirements are met. The Morlet mother wavelet basis function is used as the activation function between the input layer and the hidden layer. The gradient correction method is used to modify the weights of the network and the wavelet basis function parameters. The final network structure is determined according to the minimum prediction error.

After comparing the training results of the network, the number of nodes in the hidden layer is determined as 35, with faster global convergence speed and smaller error of the network.

(3) Results and discussion

This experiment takes five tags as a group to conduct the experiment of 3D coordinate measurement of RFID multi-tag network. Before acquiring the coordinates of RFID tags at different positions, the experiment starts the RFID tag reading distance dynamic test system to obtain the reading distance corresponding to the RFID multi-tag network distribution. The experimental process is as follows:

Five different RFID tags with brackets are placed on the turntable. According to the need of the experiment, the heights of RFID tags are adjusted so that the heights of the five RFID tags are different. The control computer controls the vertical camera to get the five RFID tags' images. After that, the denoising method is used to preprocess the obtained five tags' images. The 2D coordinates of the five RFID tags are obtained according to the method described. The vertical coordinates of the five RFID tags are obtained according to the method described. Finally, the 3D coordinates of the RFID tags are obtained.

Take different groups which each group has five different RFID tags to conduct the experiment. The complete data in Tables 5.6 and 5.7 are public and can be obtained

Table 5.6 RFID multi-tag network 3D coordinates and corresponding reading distance

x_1/m	y_1/m	z_1/m	…	x_5/m	y_5/m	z_5/m	d_r/m
0.815	0.278	0.216	…	0.957	0.592	0.235	1.23
0.906	0.547	0.189	…	0.485	0.759	0.191	1.65
0.127	0.957	0.117	…	0.800	0.655	0.170	0.96
⋮	⋮	⋮		⋮	⋮	⋮	⋮
0.913	0.962	0.165	…	0.142	0.135	0.065	1.11
0.632	0.157	0.203	…	0.426	0.849	0.158	1.36
0.097	0.970	0.079	…	0.915	0.633	0.149	1.68

Table 5.7 Wavelet neural network prediction results

Sample ID	d_r/m	d_p/m	E/%
1	1.23	1.22	0.81
2	1.65	1.67	1.21
3	0.96	0.95	1.04
	⋮	⋮	⋮
98	1.11	1.11	0
99	1.36	1.37	0.74
100	1.68	1.69	0.6

by accessing http://blog.csdn.net/zx3531675/article/details/79442705. One type of distribution structure is illustrated in Fig. 5.32.

After acquiring 3D coordinate distribution data of RFID multi-tag network and corresponding reading distance data, the wavelet neural network is used to model the nonlinear relationship between the 3D coordinate distribution of RFID multi-tag network and the corresponding RFID tag reading distance. A total of 500 groups of RFID multi-tag 3D coordinates and corresponding reading distance data are collected. In this paper, 400 groups among them are randomly selected to train the network, and the trained network is used to predict the reading distance of the rest 100 groups data. The prediction relative error is defined in Eq. (5.50).

$$E = \frac{|d_p - d_r|}{|d_r|} \times 100\% \tag{5.50}$$

In Eq. (5.50), d_p is wavelet neural network prediction value, d_r is the raw data obtained from the experiment.

The experimental prediction results of wavelet neural network are shown in Table 5.7.

When the number of nodes in the hidden layer is 35, the average prediction relative error is 0.0071. The average prediction relative error is small. The wavelet neural

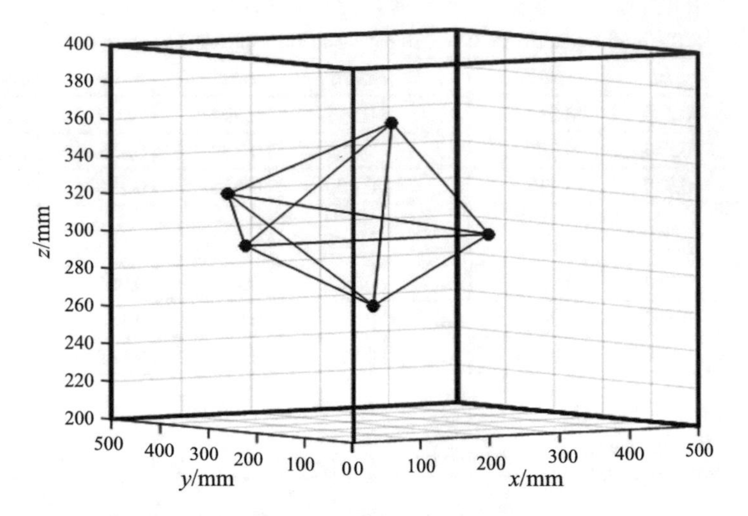

Fig. 5.32 The spatial structure of five RFID tag network

network can effectively model the nonlinear relationship between the RFID multi-tag 3D coordinate distribution and the corresponding RFID tag reading distance. The wavelet neural network model can predict the reading distance of RFID multi-tag networks effectively.

In order to verify the superiority of the wavelet neural network, the GA-BP neural network and the PSO neural network are compared with the wavelet neural network, respectively, in this paper. The time cost is calculated 200 times to get the average time cost of different networks by using the DELL computer (Inter(R) Core(TM) i5-2450M CPU @ 2.50GHz).

From Fig. 5.33, comparing with PSO neural network and GA-BP neural network, the prediction relative error of wavelet neural network is smaller. The wavelet neural network can better model the nonlinear relationship between the 3D coordinates of RFID multi-tag network and the corresponding RFID tag reading distance. From Tables 5.8 and 5.9, the average prediction relative error of PSO neural network is 2.77% and the average prediction relative error of GA-BP neural network is 3.44%. Comparing with PSO neural network and GA-BP neural network, the average prediction relative error of wavelet neural network is only 0.71%. Besides that the time cost of wavelet neural network is 2.17s. The time cost of wavelet neural network is also less than other two methods. Compared with corresponding method, the proposed method has two advantages. On the one hand, the prediction value of the proposed method is closer to the experimental value, so the relative error is smaller. On the other hand, the proposed method has better real-time performance than PSO neural network and GA-BP neural network in the detection system designed in this paper.

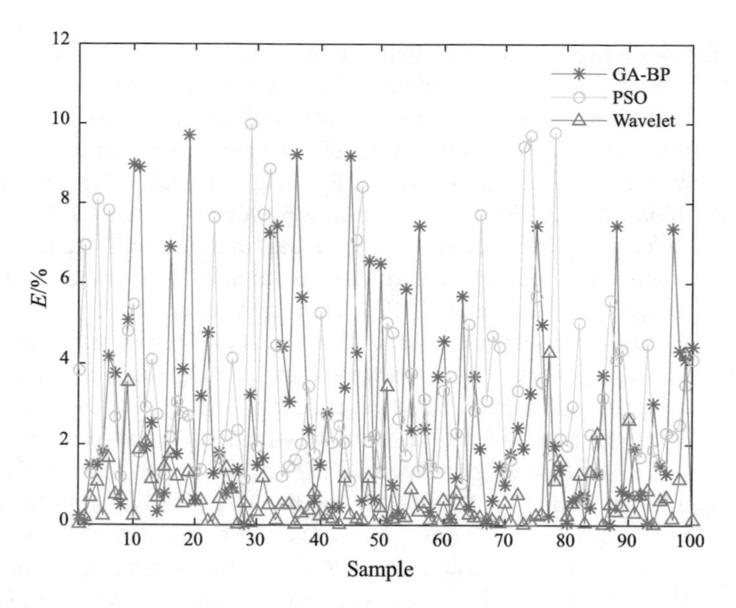

Fig. 5.33 Comparison of prediction results of different networks

Table 5.8 Average prediction relative error of different networks

Network	Average prediction relative error/%
Wavelet	0.71
PSO	2.77
GA-BP	3.44

Table 5.9 Average time cost of different networks

Network	Average time cost/s
wavelet	2.17
PSO	264.28
GA-BP	249.61

5.4 Conclusion

This chapter introduces optimization algorithm and physical anti-collision of RFID systems. First, we introduce a multi-tag optimization method based on PSO neural network. When the tags are distributed in space, their coordinates are accurately measured through a dual CCD vision system. Then, the model is established based on the relationship between the spatial distribution of the tags and the reading distance by means of PSO neural network, and the simulation of maximum reading distance is verified. The result shows that, with the help of this method, the reading distance of the RFID system has been improved. Second, the relationship between the tags' reading distance and corresponding distribution is studied via image processing and SVM. As a result, the reading distance of tags could be predicted according to the corresponding distribution. Finally, a 3D coordinate measurement system of RFID multi-tag network based on dual CCD is designed, and the 3D coordinates of RFID multi-tag network are measured by image method. When the obtained images are noisy, the wavelet threshold denoising method is used to improve the quality of the images. After that, the 3D coordinates of RFID multi-tag are obtained based on the template matching method. The measurement results show that the 3D coordinate measurement system of RFID multi-tag network is very effective and useful. Finally, the nonlinear relation model between the 3D coordinates distribution of RFID multi-tag network and the corresponding reading distance is established by using wavelet neural network. The paper defines the relative error to evaluate the predictive performance of neural networks which are discussed in this paper. Compared with GA-BP neural network and PSO neural network, the wavelet neural network can better predict the RFID multi-tag network distribution of the reading distance. The average prediction relative error is 0.71% and the average time cost of wavelet neural network is 2.17s. The time cost of the wavelet neural network is about 1% of the other two methods. The values of average prediction relative error and the time cost of the wavelet neural network are very small. Compared with corresponding method, the proposed method has two advantages.

References

1. Want R (2006) An introduction to RFID technology. Pervasive Comput 5(1):25–33
2. Valero E, Adán A, Cerrada C (2015) Evolution of RFID applications in construction: a literature review. Sensors 15:15988–16008
3. Yu Y, Yu X, Zhao Z (2016) Measurement uncertainty limit analysis of biased estimators in RFID multiple tags system. IET Meas Sci Technol 10:449–455
4. Yu Y, Yu X, Zhao Z, Wang D (2017) A novel method to evaluate the dynamic performance of RFID positioning system. J Comput 28:184–195
5. Li ZH, He CH, Li JM et al (2014) RFID reader anti-collision algorithm using adaptive hierarchical artificial immune system. Exp Syst Appl 41(5):2126–2133
6. Joo YI, Seo DH, Kim JW (2014) An efficient anti-collision protocol for fast identification of RFID tags. Wirel Personal Commun 77(1):767–775
7. Myung J, Lee W, Srivastava J (2006) Adaptive binary splitting for efficient RFID tag anti-collision. IEEE Commun Lett 10(3):144–146
8. Li BB, Feng XX, Wang CX et al (2012) Multi-passive sensors resource allocation algorithm based on information gain. Syst Eng Electron 34(3):502–507
9. Gibson R, Atkinson R, Gordon J (2016) A review of underwater stereo-image measurement for marine biology and ecology applications Oceanogr. Mar Biol 4:7257–7292
10. Ma W, Dong T, Tian H (2014) Line-scan CCD camera calibration in 2D coordinate measurement. Optik-Int J Light Electron Opt 125:4795–4798
11. Dong T, Hua D, Li Y (2014) Measuring principle of vertical target density based on single linear array CCD camera. Optik-Int J Light Electron Opt 125:176–178
12. Fahringer TW, Lynch KP, Thurow BS (2015) Volumetric particle image velocimetry with a single plenoptic camera. Meas Sci Technol 26:115–201

13. Zhou F, Wang Y, Peng B (2013) A novel way of understanding for calibrating stereo vision sensor constructed by a single camera and mirrors. Meas 46:1147–1160
14. Chen F, Chen X, Xie X (2013) 2013 Full-field 3D measurement using multi-camera digital image correlation system. Opt Laser Eng 51:1044–1052
15. Venkataraman K, Jabbi AS, Mullis RH (2014) Systems and methods for measuring depth using images captured by camera arrays: U. S. Patent. 8885059
16. Wang J, Fang K, Pang W (2017) Wind Power Interval Prediction Based on Improved PSO and BP Neural Network. J Elect Eng Technol 12:989–995
17. Ding S, Su C, Yu J (2011) An optimizing BP neural network algorithm based on genetic algorithm. Artif Intell Rev 36:153–162
18. Wang D, Luo H, Grunder O (2017) Multi-step ahead electricity price forecasting using a hybrid model based on two-layer decomposition technique and BP neural network optimized by firefly algorithm. Appl Energy 190:390–407
19. Doucoure B, Agbossou K, Cardenas A (2016) Time series prediction using artificial wavelet neural network and multi-resolution analysis: application to wind speed data. Renew Energy 92:202–211
20. Sharma V, Yang D, Walsh W (2016) Short term solar irradiance forecasting using a mixed wavelet neural network Renew Energy 90:481–492
21. Yu X, Yu Y, Wang D et al (2016) A novel temperature control system of measuring the dynamic UHF RFID reading performance. In: 2016 Sixth International Conference on instrumentation and measurement, computer, communication and control (IMCCC). IEEE, pp 322-326
22. Yu Y, Yu X, Zhao Z et al (2016) Measurement uncertainty limit analysis of biased estimators in RFID multiple tags system. IET Sci Meas Technol 10(5):449–455
23. Tsanakas JA, Chrysostomou D, Botsaris PN et al (2015) Fault diagnosis of photovoltaic modules through image processing and Canny edge detection on field thermographic measurements. Int J Sustain Energy 34(6):351–372
24. Singh S, Datar A (2015) Improved hash based approach for secure color image steganography using canny edge detection method. Int J Comput Sci Netw Secur (IJCSNS) 15(7):92–98
25. Goulart JT, Bassani RA et al (2017) Application based on the Canny edge detection algorithm for recording contractions of isolated cardiac myocytes. Comput Biol Med 81:106–110
26. Pluhacek M, Senkerik R, Davendra D et al (2013) On the behavior and performance of chaos driven PSO algorithm with inertia weight. Comput Math Appl 66(2):122–134
27. Sung WT, Chung HY (2014) A distributed energy monitoring network system based on data fusion via improved PSO. Measurement 55:362–374
28. Khanna V, Das BK, Bisht D et al (2015) A three diode model for industrial solar cells and estimation of solar cell parameters using PSO algorithm. Renew Energy 78:105–113
29. Sedghi M, Aliakbar-Golkar M, Haghifam MR (2013) Distribution network expansion considering distributed generation and storage units using modified PSO algorithm. Int J Electr Power Energy Syst 52:221–230
30. Plaza A, Martinez P, Perez R et al (2002) Spatial/spectral endmember extraction by multidimensional morphological operations. IEEE Trans Geosci Remote Sens 40(9):2025–2041
31. Shao YH, Chen WJ, Deng NY (2014) Nonparallel hyperplane support vector machine for binary classification problems. Inform Sci 263:22–35
32. Pedregosa F, Varoquaux G, Gramfort A et al (2011) Scikit-learn: mchine learning in python. J Mach Learn Res 12:2825–2830
33. Zhou Y, Yu X, Wang D (2017) Optimization analysis of distribution of RFID multi-tag based on GA-BP neural. In: IEEE 2nd international conference network advanced information technology electronic automation control

Chapter 6
Deep Learning and RFID System Physical Anti-Collision

Radio Frequency Identification (RFID) technology has significant traits such as batch reading, non-line-of-sight communication, and teletransmission, which is a non-contact identification technology automatically. As its peculiarities, RFID has a steady increase in warehouse inventory, electronic payment, object tracking, target detection, smart city, automatic driving, and so on [1–3]. What is more, the major superiority of RFID is the simultaneous identification and location of multi-target. In the field of RFID, the reading performance of tags is an important performance indicator for measuring tag. Related studies have shown that the tags' geometrical distribution has an important influence on the tags' reading performance. But when multiple tags are in the reader's recognition range and the reader responds to them at the same time, the reader will not recognize the tag correctly. This phenomenon is called tag collision. At present, we mainly solve the collision problem of multiple tags in the same channel through some algorithms such as the ALOHA algorithm and binary tree algorithm. However, these algorithms solve the conflicts in the communication protocol layer, which cannot deal with the impact of external interference on air interface communication.

In order to further analyze the influence of tag distribution on the overall performance of the tags, we designed the RFID tag distribution optimization system based on machine vision inspection. Firstly, we use CCD cameras to capture the images of the tags. Secondly, due to the degradation in the acquired images, the deep learning method is used to restore the degraded images during image acquisition. Thirdly, we use an image matching algorithm to obtain the location information of the RFID tag network. Fourthly, by using the RFID dynamic detection system, the reading distance of the tag network is measured. Finally, aiming at the nonlinear relationship between tag network distribution and corresponding reading distance, deep learning is used to model the nonlinear relationship to optimize the tag distribution.

In 2006, Hinton and his collaborators put forward the concept of deep learning [4, 5]. Deep learning methods including CNN, DBN, AE, RNN, etc., have been widely used in many fields [6–8]. At the same time, deep convolution neural network (CNN) has drawn increasing attention in machine learning [9, 10]. Due to the

X. Yu et al., *Physical Anti-Collision in RFID Systems*,
https://doi.org/10.1007/978-981-16-0835-3_6

successful applications of CNN in the fields of image feature extraction and recognition, it provides a new notion to solve the problem of image denoising [11, 12]. A completely new convolutional neural network structure applied to image processing is put forward, which can fully excavate the internal features and the prior knowledge of the image. Plain neural networks can be easily trained by random gradient descent on very large data sets. Secondly, the image processing results could be combined with the subsequent deep belief network (DBN) to obtain the relationship between the RFID tag group's 3D positions and corresponding reading distance.

In order to optimize the tags' geometrical distribution and improve the tags' reading performance, this chapter introduces deep learning to pre-process RFID multi-tag images and analyzes the reading performance of multi-tag structures in depth. Therefore, the main contents of this chapter are as follows. First, the theory related to deep learning is briefly reduced. Secondly, the constructed RFID multi-tag 3D measurement system based deep learning is introduced. Thirdly, the multi-level wavelet CNN (MWCNN) and flexible feed-forward denoising convolutional neural network (FDnCNN) are used to process images. The network denoises and analyzes the multi-label network separately. Finally, the three-dimensional coordinates of the multi-label and the corresponding reading distance are modeled by a deep belief network (DBN), and the system is analyzed and evaluated.

6.1 RFID Multi-tag 3D Measurement System

6.1.1 System Architecture

The communication between readers and RFID multi-tag shares the same wireless channel. When RFID multi-label enters the reader's reading range, the data will return to the reader simultaneously, arousing information collision. Data collision will make the readers unable to identify the useful information of the labels, leading to inaccurate label positioning, lower reading recognition efficiency, larger missed reading rate, and longer recognition delay, which seriously limits the application field of RFID. Therefore, the image processing of multi-label localization is proposed, which not only can effectively avoid collision caused by channel interference, but also can read and identify label information more quickly. It will undoubtedly be a breakthrough direction. In [13, 14], an RFID multi-tag 2D measurement system by one camera was designed to extract the position of the tags by flood filling and morphological methods, and establish the direct linear transformation (DLT) between the pixel coordinates and the actual spatial coordinates of the tags.

A 3D structure prediction system for the RFID tag group is proposed, which is to find out the optimal geometric distribution structure of RFID tag groups and further improve the reading distance of the RFID tag group. The schematic diagram of the 3D structure prediction system for the RFID tag group is shown in Fig. 6.1a The physical map is shown in Fig. 6.1b. The block diagram is shown in Fig. 6.2. The

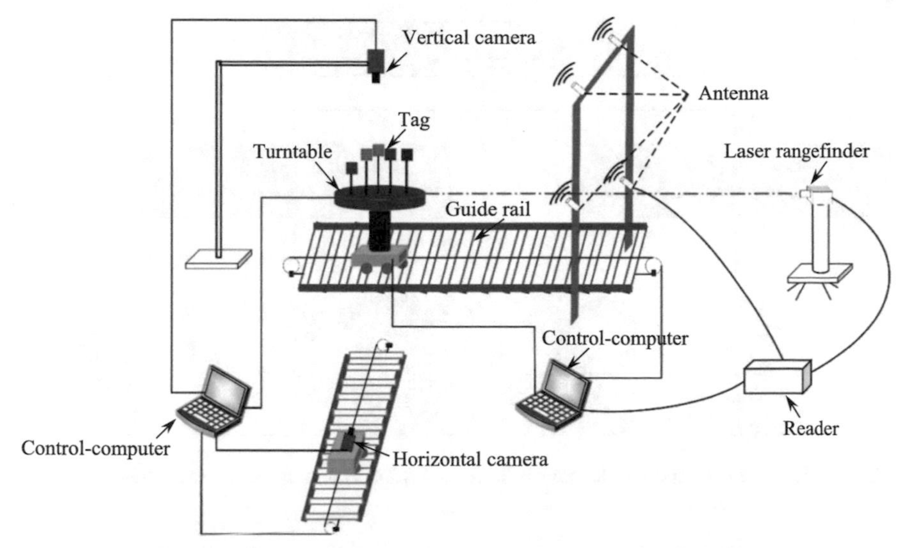

(a) Schematic diagram of 3D structure prediction system for RFID tag group

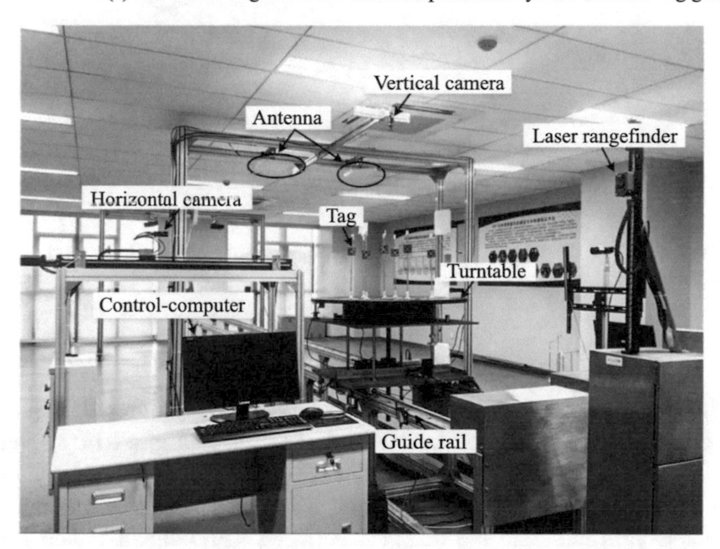

(b) Physical map of 3D structure prediction system for RFID tag group

Fig. 6.1 3D structure prediction system for RFID tag group

proposed system is based on stereovision. The system is composed of two parts. The first part is RRRS (RFID tag reading range system), which is used to obtain the reading distance of the RFID tag group. The reading distance is an important index to evaluate the performance of the RFID tag group. The RRRS is mainly composed of a control-computer, reader, antenna, laser rangefinder, RFID tag, guide rail, and so on, which is shown in Fig. 6.3. The RRRS simulates the steps of tags entering

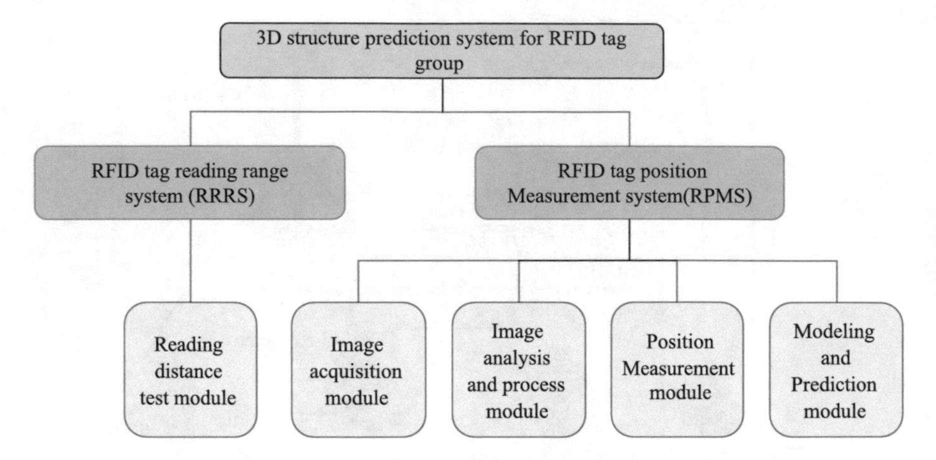

Fig. 6.2 The block diagram of the 3D structure prediction system for RFID tag group

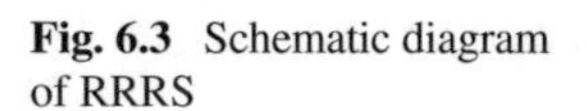

Fig. 6.3 Schematic diagram of RRRS

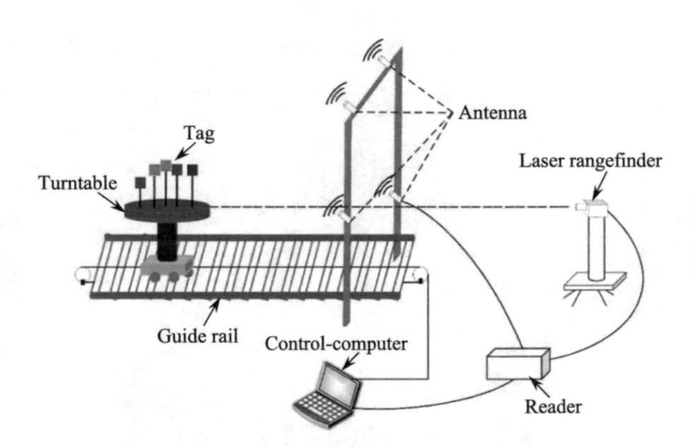

and leaving the warehouse. The turntable is placed on the guide rail. The measured tags are placed on the turntable. An RFID reader and multiple RFID antennas are mounted on the antenna frame to simulate the gate. The turntable moves in the direction of the gate. When the reader reads all the tags on the turntable, the laser rangefinder is activated, and the distance between RFID antennas and RFID tags is measured. The value of the measured distance is output as the reading distance of RFID tags. The second part is RPMS (RFID tag position measurement system), which is used to measure the RFID tag group's 3D positions by image matching method. The RPMS consists of a vertical CCD camera, horizontal CCD camera, turntable, guide rail, control-computer, and so on. After the RRRS obtains the tag group's reading distance, in order to measure the RFID tag group's 3D positions, the RPMS is started. The RPMS controls the vertical camera and horizontal camera from different angles to obtain the images of the RFID tag group. The blurs and noise in the captured images are removed by the image processing method. Then, based on image processing, the 3D positions of the RFID tag group are determined by the image matching method. Finally, DBN is used to model the nonlinear relationship between the RFID tag group's 3D positions and corresponding reading distance. The DBN can predict the reading distance of unknown tag groups and find out the optimal distribution structure of tag groups corresponding to the maximum reading distance.

The software of the 3D structure prediction system is written in C + + language. The software consists of four layers: the application interface layer, the system parameter configuration layer, the test protocol layer as well as the data storage and process layer. The relevant parameters and commands are transmitted between the layers.

The data acquisition electronics of 3D structure prediction systems are mainly composed of reader, reader antenna, laser rangefinder, and CCD camera. The RFID reader uses Impinj's Speedway Revolution R420 ultra high frequency reader. The reader antenna uses the Larid A9028 far-field antenna. The laser rangefinder uses Wenglor Company's X1TA101MHT88 laser rangefinder. The CCD camera lens uses the Japanese Utron Company's 2 million pixel level FV0622 industrial lens with a focal length of 6.5 mm.

6.1.2 Image Process Module

In the RFID tag position measurement system, an image acquisition system is devised to build a basic measurement platform. In this system, a dual camera collect the multi-label images from two different angles. When the multi-label position is measured, the perpendicular camera captures the top view of the multi-label, in which the placement on the horizontal plane can be seen. The horizontal camera captures perpendicular multi-image as the turning disc rotates clockwise, where the placement of the multi-label in the perpendicular direction can be got.

The flowchart of the RFID multi-label localization system is shown in Fig. 6.4. First, set the speed and height of the turning disc so that all the labels are fully visible to the camera and the turning disc rotates clockwise at 5 s/rad. The multi-label images are captured by a dual CCD camera and pre-processed to get clearer images. The center of the turning disc is marked. The horizontal coordinates of labels can be calculated by template matching on the perpendicular camera. With the rotation at a constant speed, the horizontal camera can take the vertical images of the multi-label at different angles, and the location of the multi-label can be computed by template matching. Finally, SAM converts the coordinates of the image pixel into multi-label coordinates of the real space. Randomly distribute multi-label again, after all procedures are completed.

In the image acquisition system, the perpendicular camera firstly acquires a multi-label image in the horizontal direction, as shown in Fig. 6.5a, in which the center position of the turntable is marked. Second, the horizontal camera acquires multi-label images in the vertical orientation as shown in Fig. 6.5b. The relative motion between the label and the camera produces motion blur and the noise from various sources caused by multi-label in optical imaging system are also the main reason of image degradation. We focus on the multi-label image denoising and deblur to

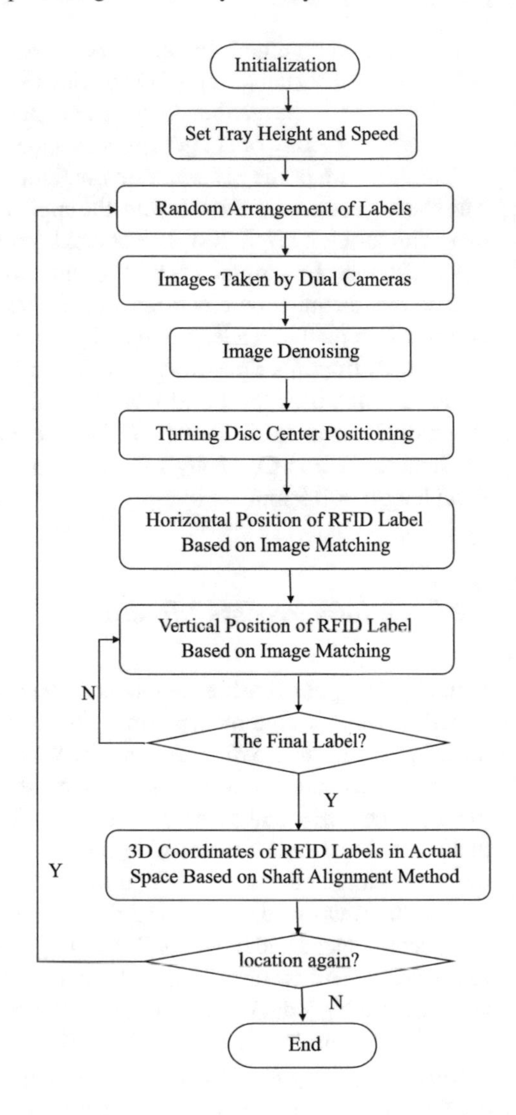

Fig. 6.4 Flow chart of multi-label localization system

improve the image quality through pre-processing and get a potentially sharp image. As can be seen from Fig. 6.5a, the multi-label image in the horizontal orientation is set to a black turning disc as the image background, and the label position of the arrow mark on the tray surface is easily extracted. The acquired image can locate multi-label completely and clearly. In Fig. 6.5b, it is more difficult to process multi-label image in the complicated background. Therefore, the multi-label image processing is only for the multi-label image of the complex background in the vertical direction.

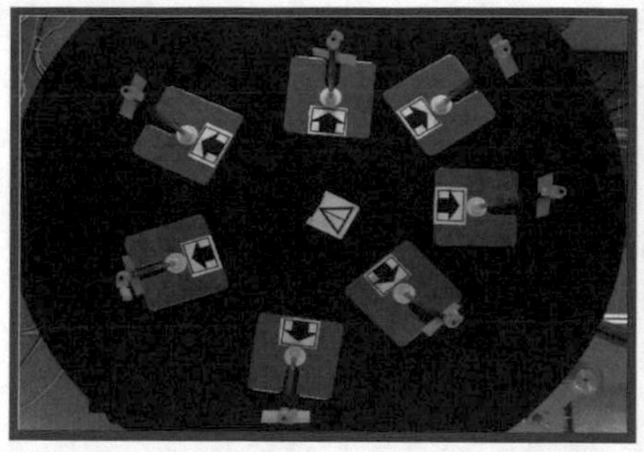

(a) Multi-label image in the horizontal direction

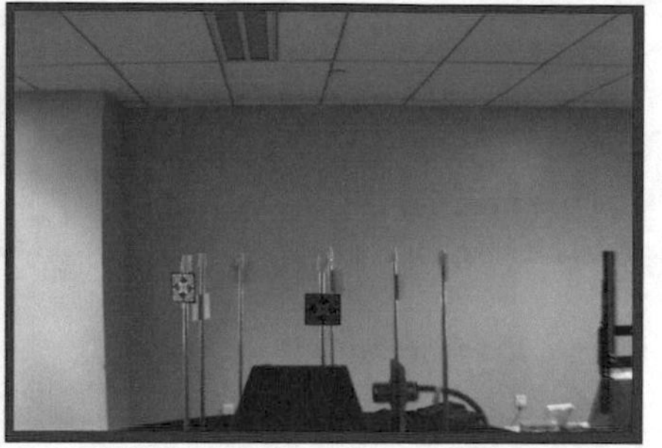

(b) Multi-label image in the vertical direction

Fig. 6.5 Multi-label image taken by the designed system

6.2 The Theory RFID Multi-tag Image Denoising by FDnCNN

The purpose of image processing is to enhance the target of interest in the image, make the area of the label clearer, and provide a basis for subsequent multi-label dynamic measurement and other operations. In multi-label image acquisition, various kinds of noises generated in the digital optical imaging system make the sharp image add extra noise, which will largely affect the precision of multi-label 3D dynamic measurement system. Gaussian white noise does not exist in the camera, but it is feasible to approximate it to additive white Gaussian noise (AWGN) [15, 16]. For AWGN, the advanced image denoising way based on the flexible deep convolutional neural network is proposed, which can get potentially tidy images. Its merit is that it can denoise AWGN with the different noise levels and spatial variability flexibly, and acquire higher quality images than other prevailing ways.

6.2.1 The Proposed FDnCNN Network Architecture

In this section, we state the proposed flexible denoising convolutional neural networks model, i. e., FDnCNN, and extend it to hand RFID multi-label image denoising tasks. Classically, building the FDnCNN network model for RFID multi-label image requires two steps: (a) Network model design, (b) Learn the network from the trained data and extend the denoising of FDnCNN to a multi-label image.

The proposed FDnCNN framework is shown in Fig. 6.6. Firstly, the original noise image y is decomposed into four down-sampled sub-images by the reversible down-sampling operation. Secondly, the four sub-images and corresponding noise level maps are used as the inputs of nonlinear mapping, and the deep convolution

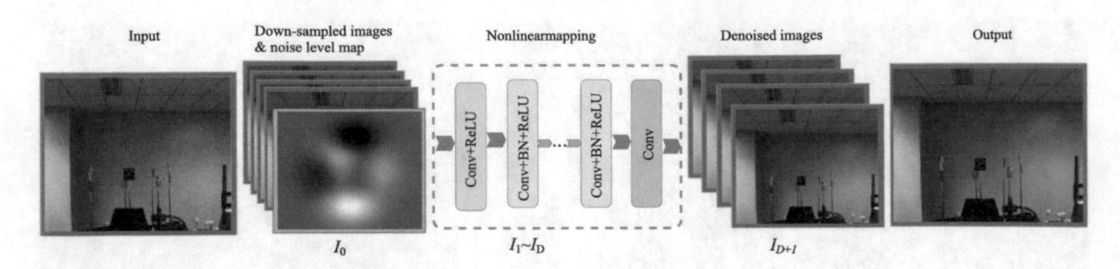

Fig. 6.6 FDnCNN network framework

neural network is D convolutional layers in the nonlinear mapping. Finally, the sub-images output from the convolution neural network is sampled up to form a new tidy multi-label image. The specific operating process of FDnCNN is as follows:

(1) Down-sampling layer

In the first layer I_0, the image y with the pixel of $h \times w \times c$ as input is shaped into four sub-sampled images with the pixel sizes of $h/2 \times w/2 \times 4c{+}1$ by Eq. (6.1).

$$I_1[c, m, n] = I\left[\left\lfloor \frac{c}{4} \right\rfloor, 2m + (c \bmod 2), 2n + \left\lfloor \frac{c}{2} \right\rfloor\right] \tag{6.1}$$

Among them, $0 \leq m < h$, $0 \leq n < w$, c refers to the number of image channels. Specifically, if the image y is a grayscale image, $c = 1$, if the image y is a color image, $c = 3$. Here, the multi-label images are all grayscale images [19]. In this study, $c = 1$. All images are sampled down by using Eq. (6.1) into four sub-images and the noise level map M to form a tensor of size $h/2 \times w/2 \times (4c + 1)$, which is used as input of the deep convolutional neural network. Figure 6.7 illustrates the image down-sampling structure. It can be seen that the original image is the noise level map added to the gray sub-images.

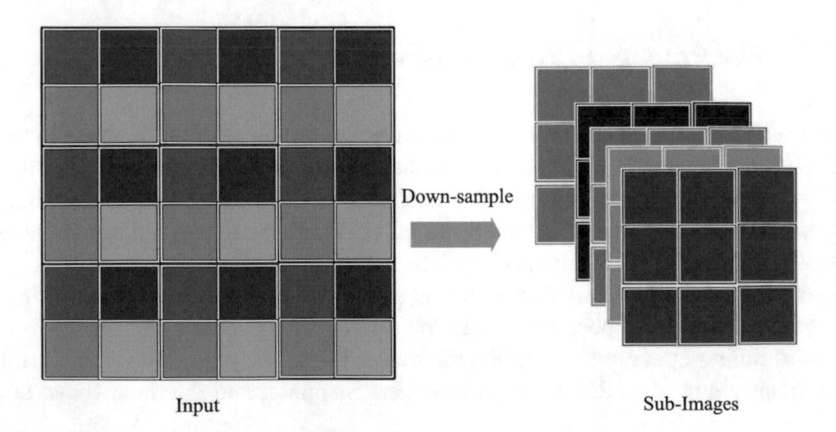

Fig. 6.7 Image reversible down-sampling structure

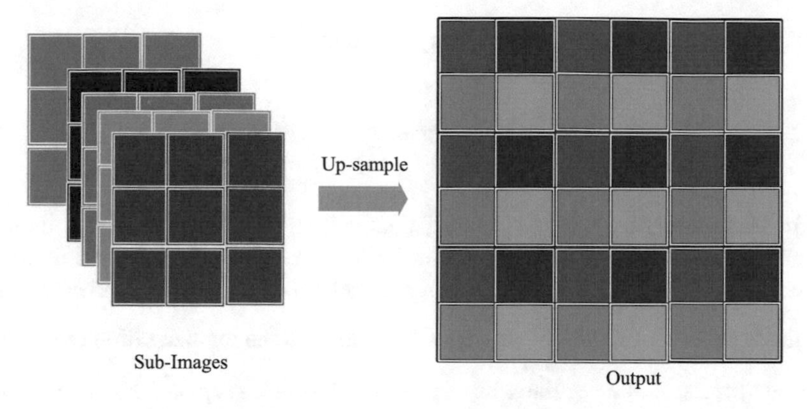

Fig. 6.8 Multi-label up-sampling structure of at upper sampling layer

(2) Non-linear mapping module

The nonlinear mapping of deep convolutional neural networks consists of *D* convolutional layers with the size of 3 × 3 after the down-sampling layer. In the nonlinear mapping, there are three types of operations: convolution (Conv), rectified linear unit (ReLU), and batch normalization (BN). In particular, the *D*-layer nonlinear mapping is characterized by: in the first layer I_1, Conv + ReLU is adopted. 64 filters with the size of $3 \times 3 \times c$ generate 64 feature maps, and the rectified linear element $\mathrm{Re}LU() = \max(, 0)$ is used for nonlinear mapping. Between the second layer I_2 and the *D-1* layer $I_{D\text{-}1}$, Conv + ReLU + BN is taken. 64 filters of size 3 × 3 × 64 are used, and batch normalization operation (BN) is added between Conv and ReLU. In I_{D}, Conv is used, in which c filters of size 3 × 3 × 64 are used to reconstruct the output.

Normally in many elementary vision images, the input and output images are required to be consistent in size, which may engender visual artifacts. In this paper, zero filling is carried out directly before each convolutional layer to ensure that the size of the image does not change with depth. Meanwhile, we can see that a simple zero-fill strategy does not produce any boundary artifacts.

(3) Upper sampling layer

At the $D + 1$ layer $I_{\mathrm{D+1}}$, sample the low-resolution output of the *D* layer up to the original image resolution as Fig. 6.8.

6.2.2 Noise Level Map

To develop the adaptability of the FDnCNN denoiser to denoise different noise levels, the model-based image denoising method is shown in Eq. (6.2) to train a discriminative solution for the following problem [9, 10].

$$\hat{x} = \min_x \Psi(y - x) + \lambda \sum_{k=1}^{K} \sum_{p=1}^{N} \rho_k((f_k * x)_p) \tag{6.2}$$

For Gaussian noise,$\Psi(z) = \frac{1}{2}\|z\|^2$, thus, $\Psi(y - x) = \frac{1}{2\sigma^2}\|y - x\|^2$ indicates the quality of data reproduction in the noise level σ. Among them, f_k*x represents the convolution of image x and k_{th} filter kernel function $f_k.\rho_k()$ expresses the k_{th} penalty function, hence, $\sum_{k=1}^{K} \sum_{p=1}^{N} \rho_k((f_k * x)_p)$ denotes the regularization term related to image priors. λ is a regularization parameter, which keeps the balance between quality of data reproduction and regularization from Eq. (6.2). If λ is too small, the first item of Eq. (6.2) plays a vital role in the estimated image to remain much noise. On the contrary, if λ is too large, the regularization item plays a decisive role, which makes the image details smooth, and the noise suppressed at the same time [11].

By optimization, Eq. (6.2) is transformed into implicit expression Eq. (6.3).

$$\hat{x} = \Re(y, \sigma, \lambda; \Theta) \tag{6.3}$$

Here, Θ is the model parameters trained. Suppose that λ is absorbed and replaced by σ, it can be rewritten as Eq. (6.4)

$$\hat{x} = \Re(y, \sigma; \Theta) \tag{6.4}$$

In other words, when the noise level σ is different, it can automatically modify the value to control the relation between the denoising effect and image particulars. It can be seen in Eq. (6.4) that FDnCNN takes the original image and noise level as inputs. Due to x and σ with different sizes, it is not possible to directly input into deep convolutional neural networks. In order to solve the problem of dimensional mismatch, σ is set for each patch, and the noise grade σ is stretched to the noise level map M. In training the network, all elements are σ in the noise level map. Equation (6.4) can be further rewritten as

$$\hat{x} = \Re(y, M; \Theta) \tag{6.5}$$

Here, the convolution structure of FDnCNN adds an estimate of the noise level in training, as shown in Fig. 6.7. The noise level is connected as a new channel added. The noise map M may be multi-channel. In the color image, the noise map M indicates the noise existing in the R, G, and B channels, respectively. Empirically, a non-uniform noise map M can signify spatially varying noise. The inputs involve the noise map M in training FDnCNN to handle various noise levels, and control the equilibrium between noise reduction and detail preservation from Eq. (6.2–6.4). Overall, the increase in the value of M enhances the noise reduction effect at the expense of removing the details of the image. In order to guarantee that M plays a role in balance control, the convolution filter is standardized by orthogonalization during training.

6.2.3 *Loss Function*

The input of FDnCNN is the noise image of RFID multi-label $y = x + v$. In image denoising, the four sub-images and a noise grade map M are selected as the input of the nonlinear network FDnCNN to continuously learn the relationship between the potential and the estimated clean image in the network.

Such internal relations are expressed by parameters in the network. In the convolutional neural network, the network parameters are adjusted through the error back-propagation algorithm in continuous learning. Assuming that the original image is x, the corresponding noise image is y, and the potential clean image is $\hat{x}$ through the networks, the relationship between the original image x and the estimated clean image can be expressed as Eq. (6.6):

$$\ell = \ell(y, x, \hat{x}; \Theta) \tag{6.6}$$

This is the loss function. The goal is to minimize the value of the loss function by continuously training the network parameters as Eq. (6.7).

$$\Theta = \min_{\Theta}(\ell) \tag{6.7}$$

In general, in a discriminative learning model, the mean square error between a potential clean image and an estimated clean image can be calculated as a loss function for learning training parameters as shown in Eq. (6.8).

$$\zeta(\Theta) = \frac{1}{2N}\sum_{i=1}^{N} \|F(y_i, M_i; \Theta) - x_i\|_2 \tag{6.8}$$

In the training, Adam is the optimal algorithm to optimize the minimization loss function. All hyper-parameters, beta1 and beta2, use the default values.

6.2.4 *Experimental Result and Real-Time Analysis*

(1) Implementation Details for Multi-label Image Denoising

In the image denoising of RFID multi-label dynamic localization system, the control-computer processor selected is Inter(R)Core(TM)i5-8300H CPU@2.3 GHz, the operating system (OS) is Windows10 64-bit, and the GPU is NVIDIA Geforce GTX 1050 Ti. All experimental results run in the software MATLAB 2016a, and the proposed FDnCNN denoising model is trained with the MatconvNet toolbox to implement convolutional neural networks (CNN) in computer vision applications. It is straightforward and efficient to train the most advanced deep convolution neural networks. Meanwhile, MatconvNet runs on the support of the compiler Visual Studio 2015. The version of the compute unified device architecture (CUDA) selects NVIDIA CUDA Toolkit 9.0, which is a general parallel computing architecture that enables the GPU to solve complex computational problems, and accelerate the operation. NVIDIA CuDNN9.1 is a GPU-accelerated library for deep neural networks, which enables deep learning to calculate on the GPU. The MatconvNet package supports software operation.

In the multi-label image denoiser based on FDnCNN, set the network depth to 15, and the minimum loss function (Eq. (6.8)) is optimized by Adam. A small batch size of 128 is selected. The learning rate of Adam algorithm starts from 10^{-3}, until it is reduced to 10^{-4}. At the time, the training error stops decreasing. The small learning rate is 10^{-6}. The FDnCNN fine-tunes in an additional 50 iterations.

(2) Training and Testing Data Set

When multi-label images are denoised, the proposed method FDnCNN demands to be trained firstly. We prepare an input–output pair of training data set $\{(y_i, M_i; x_i)\}_{i=1}^{N}$, in which, M_i is the noise level map. Forming the corresponding noise map M (in this case, it is constant and all elements areσ) is added to the tidy patch x_i by the AWGN of σ[0, 100]. We used 400 Berkeley Segmentation Dataset (BSD) images and Set 68 as training data sets, randomly tailored $N = 128 \times 8000$ patches for training, and the patch size is set to 70×70 for grayscale images. The trained FDnCNN is embedded in the control computer for receiving and handling RFID multi-label images acquired by CCD. Thus, we used the multi-label image captured from the experiment as test data sets.

(3) nComparisons to State-of-the-Arts for AWGN denoising

Due to its high flexibility, FDnCNN proposed in this paper to denoise image with AWGN can ensure the compromise between image particular and denoising. In this paper, PSNR (Peak Signal-to-Noise Ratio) and SSIM (Structural Similarity Index), as two indexes of image quality evaluation, evaluates multi-label images at different noise levels. At the different noise levels $\sigma = 10, 30, 50, 70$, the denoising results of the proposed method are shown in Fig. 6.9.

It can be seen from Fig. 6.9:

(1) In a wide range of noise levels, FDnCNN can not only effectively denoise, but also extract the edge portion of the labels from the visual perspective.
(2) Since the structure of the label images obtained by this system is relatively simpler and has no complicated image particular, the value of PSNR is comparatively higher.

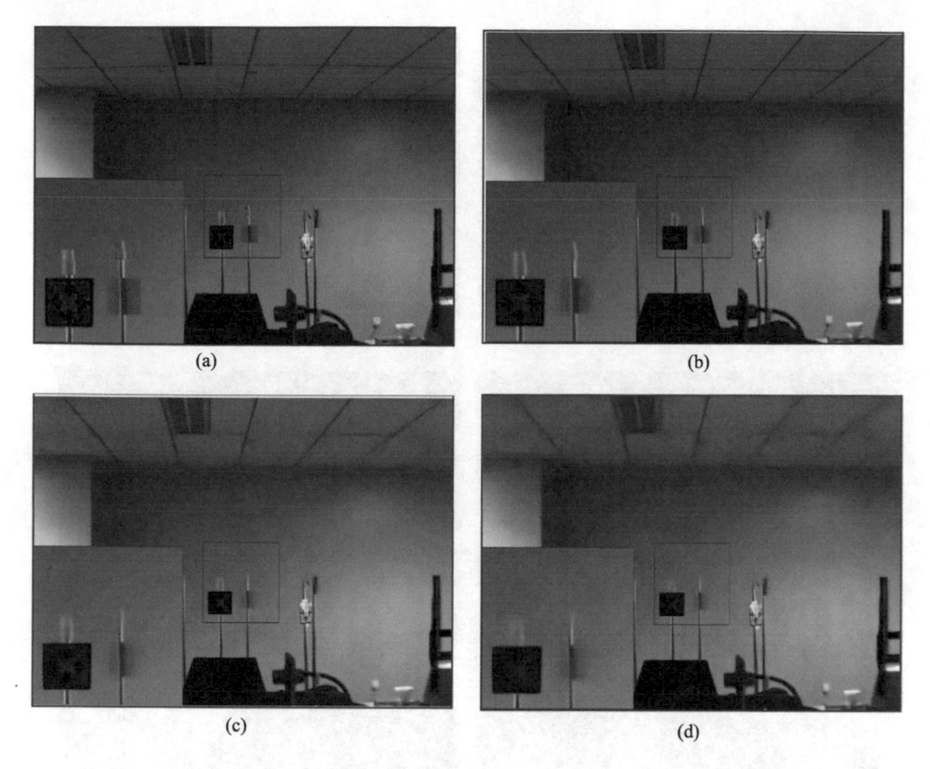

Fig. 6.9 FDnCNN denoising for different noise level **a** $\sigma = 10$ PSNR = 44.22; **b** $\sigma = 30$ PSNR = 40.89; **c** $\sigma = 50$ PSNR = 37.82; **d** $\sigma = 70$ PSNR = 37.82

(3) For the noise grade $\sigma\epsilon[0,100]$, the noise level has little effect on the image denoising results. Obviously, with the increase in noise level, clearer image edges can be extracted, but the label's details are weakened.

In Table 6.1, comparing FDnCNN with the current advanced methods (DnCNN [18] and WNNM [17]), PSNR and SSIM are the averages of them in every multi-label images collected by CCD for 50 times. When $\sigma = 20$, 30, different denoising techniques denoise multi-label images by a horizontal camera as shown in Fig. 6.10.

Referring to Table 6.1 and Fig. 6.10, it can be seen that.

Table 6.1 For different noise level maps, the images denoising quality evaluation

σ	WNNM	DnCNN	FDnCNN
10	43.0547/0.9850	42.1617/0.9718	45.0542/0.9803
20	38.6890/0.9668	42.2401/0.9736	42.7822/09,762
25	37.6780/0.9622	40.9402/0.9690	41.8047/0.9740
30	36.5203/0.9525	38.7932/0.9620	40.8968/0.9716
40	34.6586/0.9310	37.2450/0.9485	39.2790/0.9661

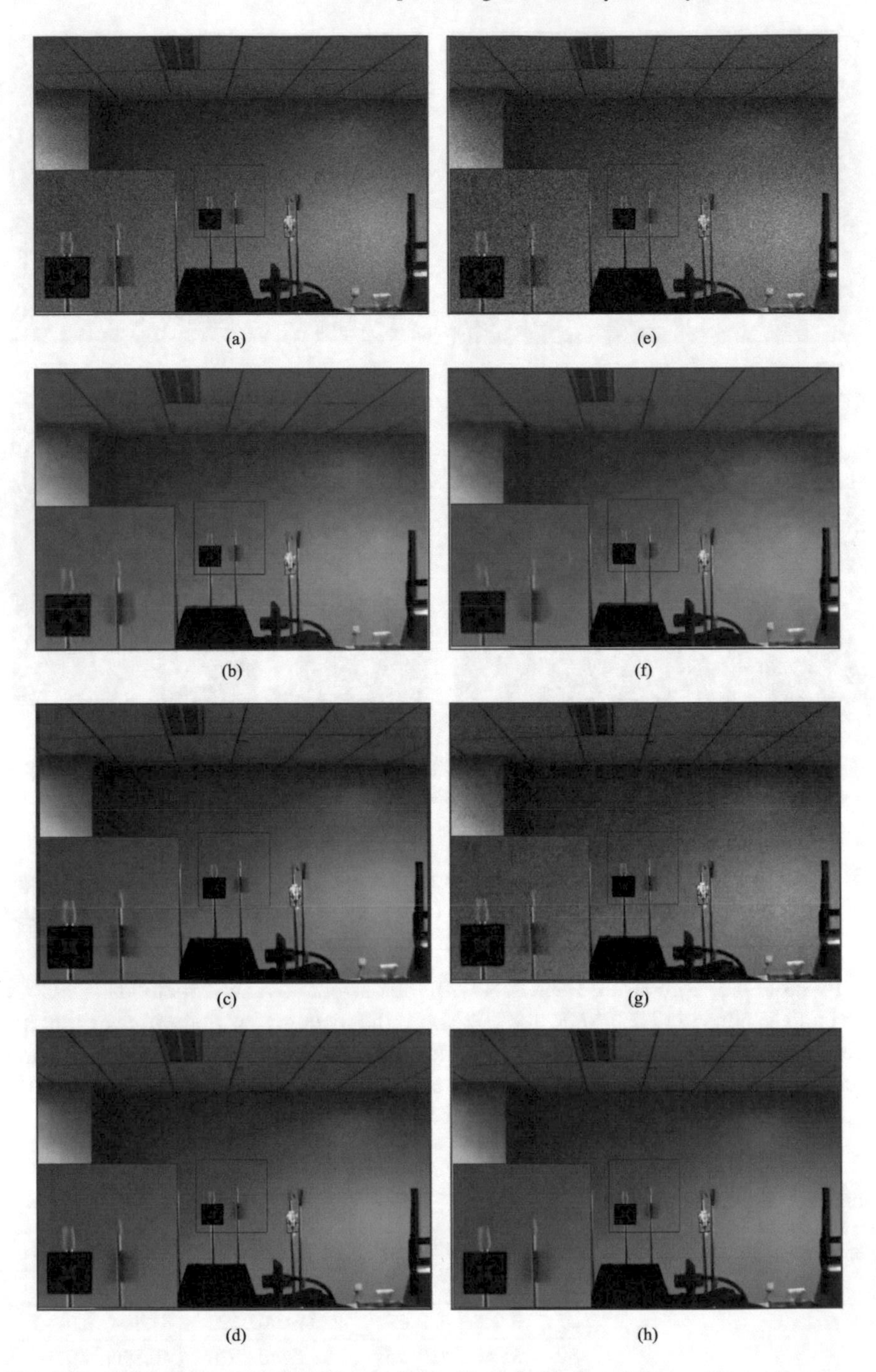

Fig. 6.10 Comparison of the denoising methods. The left side is as follows: **a** Noisy image $\sigma = 20$; **b** WNNM PSNR = 39.7230; **c** DnCNN PSNR = 42.2401; **d** FDnCNN PSNR = 42.7822; The right side is as follows: **e** Noisy image $\sigma = 35$; **f** WNNM PSNR = 36.4589; **g** DnCNN PSNR = 38.0746; **h** FDnCNN PSNR = 40.2685

(1) For the same σ, FDnCNN has the highest value of PSNR and SSIM. Hence, FDnCNN has a better image denoising effect. When σ is small, the denoising effect of FDnCNN, WNNM, and DnCNN is little different, but FDnCNN should be at least 0.5 dB higher than others.
(2) As σ increases for the same method, the PSNR of FDnCNN decreases the slowest, and the others decrease rapidly. Therefore, it can be seen that FDnCNN has ascendant denoising stability obviously, and it achieves excellent results in image denoising within a large noise grade range.
(3) At $\sigma = 20, 30$, it can be seen that the estimated clean images of FDnCNN can achieve visual effects. WNNM denoising removes the image noise and image details simultaneously. Compared with WNNM, DnCNN can get a favorable denoising effect, and smooth the image. However, compared with the two ways, FDnCNN can eliminate the noise in the flat areas, smooth the image structure, and enhance the details of the multi-label edges. Consequently, the proposed method is more conducive to multi-label localization and has the strongest perception ability.
(4) Denoising of spatially variant AWGN

FDnCNN in handling spatially variant AWGN shows its flexibility. The spatially variant AWGN is taken by AWGN with unit standard deviation is multiplied by the noise level map with random Gauss variation, as shown in Fig. 6.11d. For the sake of demonstrating the effectiveness of FDnCNN for spatially variant AWGN, compare it with AWGN with a uniform noise level map, as shown in Fig. 6.11.

As we can see from Fig. 6.11, when dealing with AWGN with uniform noise, it cannot remove the noise in the stronger area, but it can smooth the details of the area with a lower noise level. When dealing with spatially variant AWGN, FDnCNN has strong flexibility, and gets a good visual effect for different noise level areas.

(5) Real-time Performance

In addition to the analysis of visual quality and image quantification, a crucial indicator of evaluation is the speed of image denoising. All evaluation indicators are performed in the MATLAB 2016a environment. In Table 6.2, real-time comparisons are made between WNNM, DnCNN, and FDnCNN, and the time shown is the average time required for the corresponding procedures that run 50 times. The time of WNNM is the average time to handle an image with a noise level of 10. If the noise level is higher, WNNM takes a longer time. Besides, the cuDNN deep learning library can speed up the calculation of DnCNN and FDnCNN. It takes about two days to train the DnCNN network model, and FDnCNN was trained for about three days. After the training is completed, the already trained model can always handle the image denoising. For DnCNN, Table 6.2 represents the time required for 17 convolution layers to deal with a multi-label image. FDnCNN is the time required for 15 convolution layers to denoise.

If we don't take the CPU and GPU memory transfer time into account, both DnCNN and FDnCNN will show breathtaking dominant positions over other

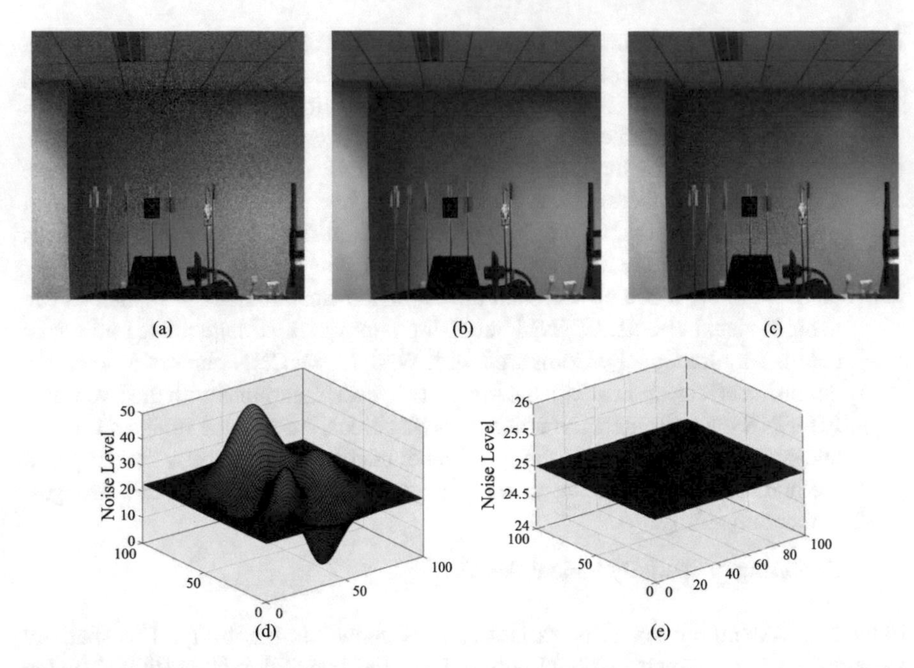

Fig. 6.11 Denoising results of spatially variant noise **a** Image with spatially variant noise; **b** Image denoising of AWGN with Gauss; **c** Image denoising of AWGN with uniform noise map; **d** AWGN distribution with Gauss; **e** AWGN distribution with uniform noise map

Table 6.2 The real-time performance of WNNM, DnCNN, FDnCNN

Method	WNNM	DnCNN	FDnCNN
Time/*s*	282.9	0.1246	0.048

methods. WNNM spends more time in denoising because WNNM in the CPU needs to iteratively calculate. DnCNN and FDnCNN get benefits from the processing power of the GPU, so they accomplish faster than WNNM. Moreover, FDnCNN is faster and more flexible than DnCNN. In addition, FDnCNN is almost the same in speed and effect in AWGN and spatial variant AWGN. More importantly, FDnCNN is very competitive in the sphere of current denoising algorithms, and is more suitable for improving image quality in real time.

6.3 Image Motion Blur Removal

The image motion blur removal is realized in the image analysis and process module in RPMS. Before the image motion blur removal, the image acquisition module in RPMS is started to obtain the images of RFID tags. However, in order to improve the real-time performance of RPMS, the horizontal camera acquires the images while the turntable is rotating. So, there will be a certain degree of relative motion between the horizontal camera and RFID tags, which will result in motion blurs in the acquired images. Because the tags' positions are measured by the image matching method, the motion blurs existing in the acquired images will seriously affect the accuracy of image matching, which will further affect the measurement accuracy of RFID tags' 3D positions. Therefore, in order to achieve high-precision tag position measurement, the motion blurs in the acquired images must be removed. In this paper, we propose the knife-edge and Wiener filtering method to restore the degraded images. First, the knife-edge method is applied to reckon the PSF, which is also called the motion blur kernel of the degraded images. Then, we use the Wiener filtering method to restore the degraded images.

There is a certain angle between the direction of the knife-edge and the image sampling. In order to obtain an accurate edge spread function (ESF), the region near the knife-edge must be over-sampled and fitted into the ESF(x) by interpolation. The ESF(x) is derived to get the line spread function LSF(x), which is shown in Eq. (6.9).

$$\mathrm{LSF}(x) = \frac{\mathrm{dESF}(x)}{\mathrm{d}x} \tag{6.9}$$

Fourier transform operation is performed on LSF and we get MTF. The equation is shown in Eq. (6.10).

$$\mathrm{MTF}(\xi) = \int_{-\infty}^{+\infty} \mathrm{LSF}(x)\exp(-\mathrm{i}2\pi\xi x)\mathrm{d}x \tag{6.10}$$

After that, the MTF is derived to obtain the PSF. The equation of the PSF is shown in Eq. (6.11). The results are shown in Fig. 6.12.

$$\mathrm{PSF} = \frac{\mathrm{dMTF}(f)}{\mathrm{d}f} \tag{6.11}$$

After obtaining the PSF of degraded images, we use the Wiener filtering method to recover the image [22]. Wiener filtering can handle images that are degraded by degradation functions and noise pollution [33]. The Wiener filtering method is to seek an estimate $\hat{g}$ of the uncontaminated $g(x, y)$, which is shown in (6.12):

$$e^2 = E\{(g - \hat{g})^2\} \tag{6.12}$$

In Eq. (6.12), $E\{.\}$ is the expected value of the parameters. If the image and the noise are not correlated, the minimum value of the mean squared error function in the frequency domain is as follows:

$$\hat{G}(m, n) = \left[\frac{D^*(m, n)T_f(m, n)}{T_f(m, n)|D(m, n)|^2 + T_\eta(m, n)}\right]F(m, n)$$

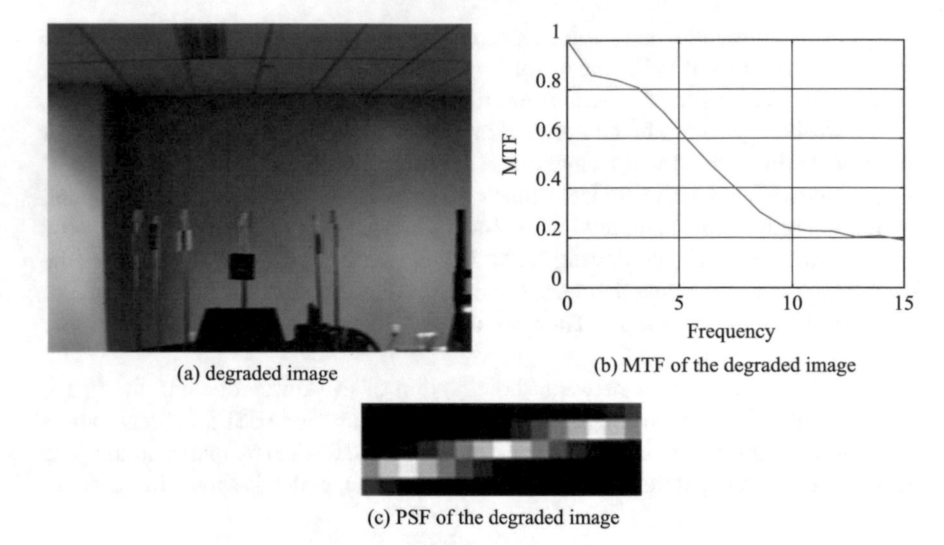

(a) degraded image

(b) MTF of the degraded image

(c) PSF of the degraded image

Fig. 6.12 Obtaining procedure of PSF

$$
\begin{aligned}
&= \left[\frac{D^*(m,n)}{|D(m,n)|^2 + T_\eta(m,n)/T_f(m,n)} \right] F(m,n) \\
&= \left[\frac{1}{D(m,n)} \frac{|D(u,v)|^2}{|D(m,n)|^2 + T_\eta(m,n)/T_f(m,n)} \right] F(m,n)
\end{aligned}
\tag{6.13}
$$

In Eq. (6.13), $D(m,n)$ is the degraded function in the Fourier field. $D^*(m,n)$ is the complex conjugate of $D(m,n)$. $|D(m,n)|^2 = D(m,n)D^*(m,n)$. $T_\eta(m,n)$=$|N(m,n)|^2$. $T_\eta(m,n)$ is the noise's power spectrum. $T_g(m,n) = |G(m,n)|^2$. $T_g(m,n)$ is the uncontaminated image's power spectrum. $F(m,n)$ is the Fourier transform of the degraded image.

From the above equations, it can be found that if there is no noise, $T_\eta(m,n) = 0$ and Wiener filtering is replaced as direct inverse filtering. If there is noise, then how to estimate $T_\eta(m,n)$ and $T_g(m,n)$ will be problematic. In practical applications, it is assumed that the degradation function is known. If the noise is Gaussian white noise, $T_\eta(m,n)$ is a certain value. However, $T_g(m,n)$ is usually difficult to estimate. An approximate solution is to use a coefficient K to replace $T_\eta(m,n)/T_g(m,n)$, so the Eq. (6.13) can be rewritten as

$$
\hat{G}(m,n) = \left[\frac{1}{D(m,n)} \frac{|D(m,n)|^2}{|D(m,n)|^2 + K} \right] F(m,n) \tag{6.14}
$$

Based on experience, the value of K is appropriately selected according to the effect of the processing. In this paper, after many experiments, the value of K is selected as 0.01. The experimental deblurring results are as follows:

Table 6.3 Results of different deblurring methods

Deblurring method	SNR/dB
Experimental initial image	15.5170
Wiener filtering method	54.1355
Constrained least square method	13.1889
Lagrange operator method	13.1636
Lucy-Richardson method	15.5645

In order to evaluate the deblurred image, this paper uses SNR. The equation of SNR is as follows:

$$\mathrm{SNR} = 10\lg\left[\frac{\sum_{i=1}^{M}\sum_{j=1}^{N} g(i,j)^2}{\sum_{i=1}^{M}\sum_{j=1}^{N} [g(i,j) - f(i,j)]^2}\right] \tag{6.15}$$

In Eq. (6.15), M is the number of pixels in the length direction of the image. N is the number of pixels in the image's width direction. $f(i, j)$ is the gray value of position (i,j) in the original image. $g(i, j)$ is the gray value of position (i,j) in the deblurred image.

For evaluating the method presented in this paper, the deblurring recovery effect of the presented method is compared with other methods, which are all traditional restoration algorithms. The results of the experiment are shown in Table 6.3.

From Table 6.3 and Fig. 6.13, we can draw the conclusion that the motion blurs in the images have been well removed. In addition, Wiener filtering method has the best deblurring effect compared with other methods and the deblurred image has the highest value of SNR. Therefore, the Wiener filtering method will be utilized in this article to restore the degraded images.

6.4 Multi-level Wavelet CNN for Image Restoration in Pre-Processing Sub-System

6.4.1 Image Restoration Based on Denoising Prior

At present, some literature have combined the denoising priori with model-based methods to process images. In [23], the authors exploit the recently introduced Plug-and-Play Prior approach to deal with denoising and super-resolution. In [24], a Plug-and-Play ADMM algorithm with provable fixed-point convergence is proposed. The authors compare Plug-and-Play ADMM with state-of-the-art algorithms in each problem type and demonstrate promising experimental results of the algorithm. In

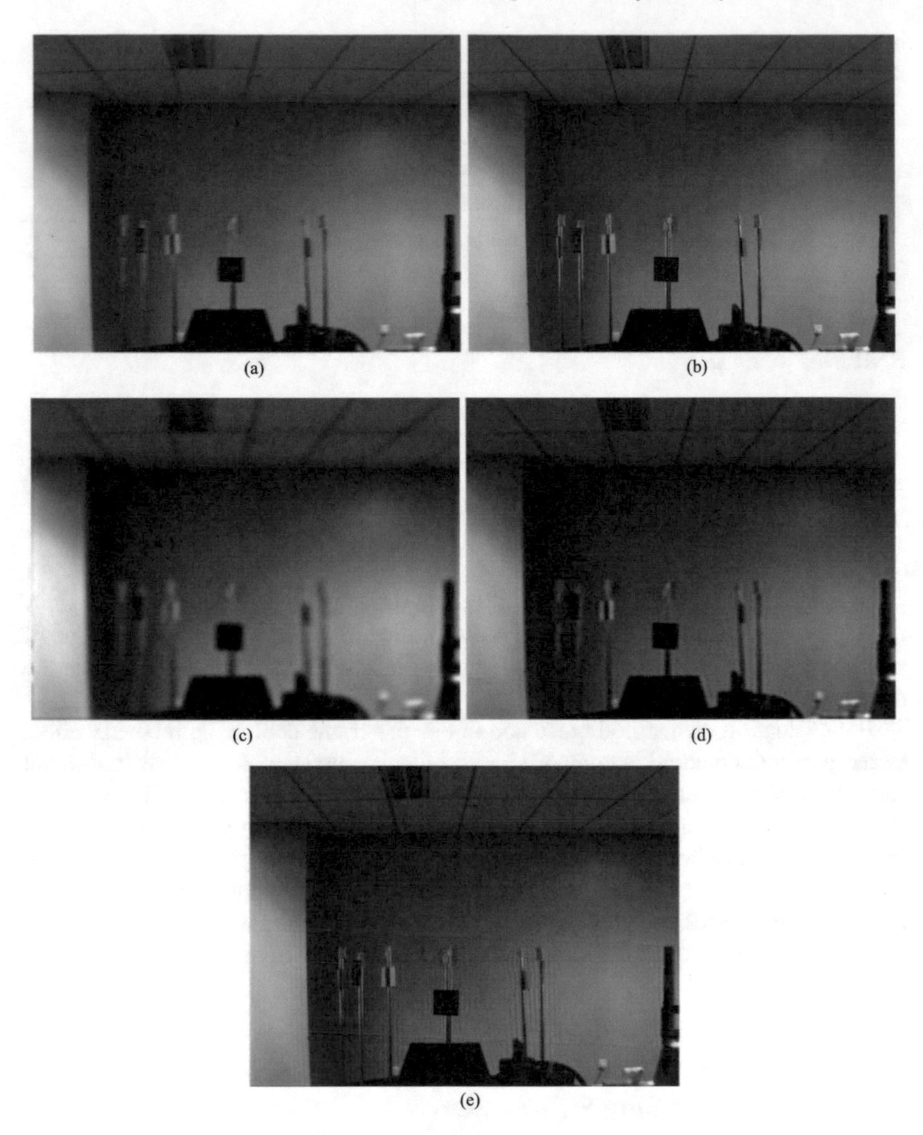

Fig. 6.13 Motion deblurred results. **a** Blurred image; **b** Restoration result of Wiener filtering method; **c** Restoration result of constrained least square method; **d** Restoration result of Lagrange operator method; **e** Restoration result of Lucy-Richardson method

[25], the authors proposed regularization by denoising (RED) method which uses the denoising engine in defining the regularization of the inverse problem. The proposed RED can incorporate any image denoising algorithm in the image deblurring and super-resolution problems. In [26], a similar plug-and-play concept was also mentioned, in which a half quadratic splitting (HQS) method was proposed for image denoising, deblurring, and repairing. In [27], the author used an alternative method

of ADMM and HQS [28] to solve the fidelity term and the regularization term. All of the above methods show that the decoupling of the fidelity term and the regularization term enables various existing models to solve different image restoration tasks.

6.4.2 *Half Quadratic Splitting (HQS)*

Generally speaking, the fidelity term and regularization term are decoupled by the variable splitting technique to plug the denoiser prior to Eq. (2)'s optimization procedure. In the HQS method, the variable z is introduced to rewritten Eq. (6.16) as

$$\hat{x} = \arg\min_{x} \frac{1}{2}\|y - \mathrm{Hx}\|^2 + \lambda\Phi(z) \quad \text{s.t.} \;\; z = x \tag{6.16}$$

Then, the HQS approach tries to solve the following problems:

$$\ell_{\mu}(x, z) = \frac{1}{2}\|y - \mathrm{Hx}\|^2 + \lambda\Phi(z) + \frac{\mu}{2}\|z - x\|^2 \tag{6.17}$$

In Eq. (6.17), μ is the penalty factor, which changes repeatedly in non-descending order. Equation (5) can be solved by the following iterations:

$$\begin{cases} x_{k+1} = \arg\min_{x} \|y - \mathrm{Hx}\|^2 + \mu\|x - z_k\|^2 \\ z_{k+1} = \arg\min_{z} \frac{\mu}{2}\|z - x_{k+1}\|^2 + \lambda\Phi(z) \end{cases} \tag{6.18}$$

It can be seen that the fidelity and regularization terms are decoupled into two separate sub-problems. Specifically, the fidelity term corresponds to the quadratic regularized least squares problem, which can quickly solve different degenerate matrices. The direct solution can be obtained by

$$x_{k+1} = (H^{\mathrm{T}}H + \mu I)^{-1}(H^{\mathrm{T}}y + \mu z_k) \tag{6.19}$$

The regularization term involving Eq. (6b) can be rewritten as

$$z_{k+1} = \arg\min_{x} \frac{1}{2(\sqrt{\lambda/\mu})^2}\|x_{k+1} - z\|^2 + \Phi(z) \tag{6.20}$$

By the Gaussian denoiser, Eq. (8) can denoise noisy images with different noise levels. From the above analysis, we can see that any Gauss denoiser can be integrated into Eq. (6) to solve the image inverse problem. Furtherly, we rewrite Eq. (6.20) as

$$z_{k+1} = \mathrm{Denoise}(x_{k+1}, \sqrt{\lambda/\mu}) \tag{6.21}$$

By Eqs. (6.20) and (6.21), we can conclude that the image priori can be replaced by the denoiser priori. This excellent feature has the following advantages. Firstly, the various image inverse problems can be solved by using different gray or color denoiser. Secondly, when solving Eq. (6.16), the explicit image priori $\Phi(\bullet)$ may be unknown. Thirdly, several different image prior denoisers can be combined to solve a specific problem.

6.4.3 Method from Multi-level WPT to MWCNN

In this chapter, we use multi-level wavelet convolution neural network to train the denoiser priori. First, we introduce multi-level wavelet packet transform (WPT). Then we present our MWCNN motivated by WPT, and describe its network architecture.

In the case of the two-dimensional discrete wavelet transform, image x is convoluted with four sub-filters f_{LL}, f_{LH}, $f_{HL,}$ and f_{HH}, and the convolution results are sampled down to obtain the processed images x_1, x_2, $x_{3,}$ and x_4. For example, x_1 is defined as $(f_{LL} \otimes x) \downarrow_2$. Because of the biorthogonal property of wavelet transform, the original image can be reconstructed precisely by inverse transform, that is, $x = \text{IWT}(x_1, x_2, x_3, x_4)$.

In the multi-level wavelet packet transform, the sub-band images x_1, x_2, $x_{3,}$ and x_4 of two-dimensional decomposition are further decomposed. Taking two-stage wavelet packet transform as an example, the upper decomposition sub-band image x_i (i = 1,2,3or4) is further decomposed as $x_{\text{i},1}$, $x_{\text{i},2}$, $x_{\text{i},3}$, and $x_{\text{i},4}$. For the third or higher level wavelet packet transform, the situation is the same. In fact, WPT is a linear special case of FCN. In the decomposition stage, each sub-band image is convoluted and down-sampled successively. In the reconstruction stage, the sub-band image is first sampled and then deconvoluted. Finally, the original image x can be accurately reconstructed by inverse WPT.

In image inverse problems, such as image denoising, restoration, and reconstruction, some nonlinear operations, such as normalization and quantization, are usually added to deal with the results of the wavelet transform. These operations can be regarded as some kind of nonlinearity designed for a specific inverse problem. Specifically, we plan to add a CNN module between any two levels of DWT to extend WPT to multi-level wavelet CNN. In this way, the self-contained image of each level of wavelet transform can be used as the input of the CNN module, so as to make full use of the excellent time–frequency characteristics of DWT and the powerful nonlinear feature extraction ability of CNN. Compared with the traditional CNN, the time–frequency characteristics of DWT are more conducive to the preservation of high-frequency information such as image texture details.

6.4.4 Network Architecture

The key of network structure design in this paper is the design of CNN after DWT. In this paper, we use a four-layer FCN without a pool as the CNN module. At the same time, CNN is deployed in high-frequency and low-frequency sub-bands of different layers. After DWT transform, there is a certain correlation between different sub-bands. Therefore, CNN after DWT can make full use of the nonlinear characteristics of CNN to extract effective features, and suppress the correlation between each sub-band. In this paper, the CNN sub-block consists of three parts: 3*3 convolutional filter (Conv), batch normalization (BN), and rectifier linear unit (ReLU). For the last layer of CNN, only Conv term is retained to predict the residual image.

The overall structure of MWCNN in this paper is shown in Fig. 6.14. It consists of a contracted subnet and an extended subnet. In this paper, MWCNN improves the traditional network from three aspects. Firstly, DWT and IWT are used to replace the maximum pool and convolution of traditional U-Net in the up-sampling and down-sampling links. Secondly, MWCNN deploys other CNN blocks to solve the problem of increasing feature mapping channels caused by down-sampling. However, in the traditional U-Net, the feature mapping channel is not affected by the lower adoption. Thirdly, the sum of elements in MWCNN is used to combine feature graphs from contracted subnets and extended subnets. Above the improvements, the final network of this paper consists of 24 layers. Figure 6.14 shows more information about the network in this article. In this paper, MWCNN uses the Haar wavelet as the default value.

The set of all parameters in MWCNN is represented by Θ. $F(y; \Theta)$ represents the output of the network. $\{(y_i, x_i)\}_{i=1}^{N}$ represents a training set of the network. Where y_i denotes output and x_i denotes input. The cost function of MWCNN is as follows:

$$L(\Theta) = \frac{1}{2N} \sum_{i=1}^{N} \| F(y_i; \Theta) - x_i \|_F^2 \tag{6.22}$$

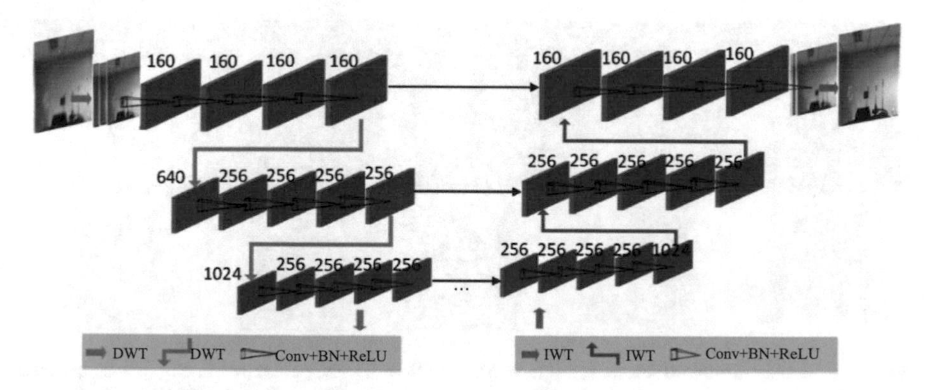

Fig. 6.14 The architecture of MWCNN

We use the ADAM algorithm to optimize the cost function to minimize the cost function.

We use four data sets to train MWCNN. The four datasets are Berkeley segmentation dataset (BSD) [11], DIV2K [15], Waterloo Exploration Database (WED) [16], and the real degraded image set generated in the actual experimental scene. We randomly extract 350 images from BSD, 1000 images from DIV2K, and 6000 images from WED. For these randomly extracted images, we add motion blurs with different blur lengths and angles to the images to generate simulated degraded images. In addition, 500 degraded real images are selected from the data set of the actual scene.

The MWCNN model is learned for each degraded setting. The network parameters are initialized based on the method described in [29]. We use the ADAM algorithm [29] with α=0.01, β_1=0.9, β_2 = 0.999, and $\in$ =10^{-8} for optimizing a mini-batch size of 24. For other superparameters of ADAM, the default settings are taken. During 40 epochs, the learning rate decays exponentially from 0.001 to 0.0001. We train our MWCNN using the MatConvNet software package with cuDNN 9.0. All experiments are performed in a MATLAB (R2016a) environment running on a PC with an Intel(R) Core(TM) i7-7700HQ CPU 2.81 GHz and NVIDIA GTX 1050 GPU.

Since the images acquired by industrial cameras are generally grayscale images, the network of this paper is only for grayscale images. After obtaining the denoiser priors by MWCNN, the acquired denoiser priors are inserted into the model-based Wiener filter to recover the image. The restored results are evaluated using PSNR and SSIM. The experimental results are shown in Fig. 6.15 and Table 6.4:

As we can see from Figs. 6.16, 6.17 and Table 6.4, the proposed MWCNN can effectively restore the degraded image. The PSNR and SSIM of the restored images are almost the same as the GSR-based method. The simulation results show that the proposed method can effectively restore the degraded images of RFID tags, and the restoration effect can meet the needs of subsequent image applications. In order to further verify the effectiveness and practicability of the proposed method, the actual degraded image is restored by this method. The restored results are as follows:

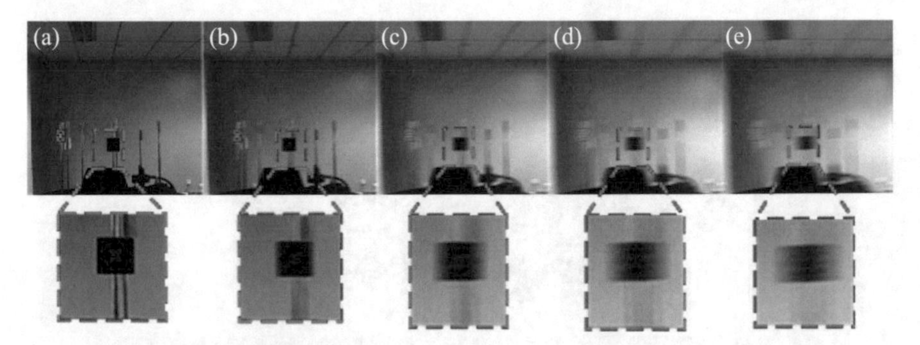

Fig. 6.15 **a** clean image **b** len = 5, angle = 5° (PSNR = 29.956) **c** len = 10, angle = 5° (PSNR = 27.493), **d** len = 15, angle = 5° (PSNR = 26.400), **f** len = 20, angle = 5° (PSNR = 25.618)

Table 6.4 Restoration results of different methods (PSNR/SSIM)

Len, angle	GSR [30]	Fast TV [31]	Ours Proposed method
len = 5,angle = 5°	35.5431/0.8502	34.1434/0.8331	36.0341/0.9634
len = 10,angle = 5°	36.6865/0.9460	33.1515/0.9043	35.8454/0.9599
len = 15,angle = 5°	36.6349/0.9458	32.0415/0.8497	35.9286/0.8820
len = 20,angle = 5°	34.4192/0.9304	30.5841/0.8686	34.0825/0.8699
len = 10,len = 10°	36.4497/0.9322	32.5135/0.8561	35.9137/0.9069

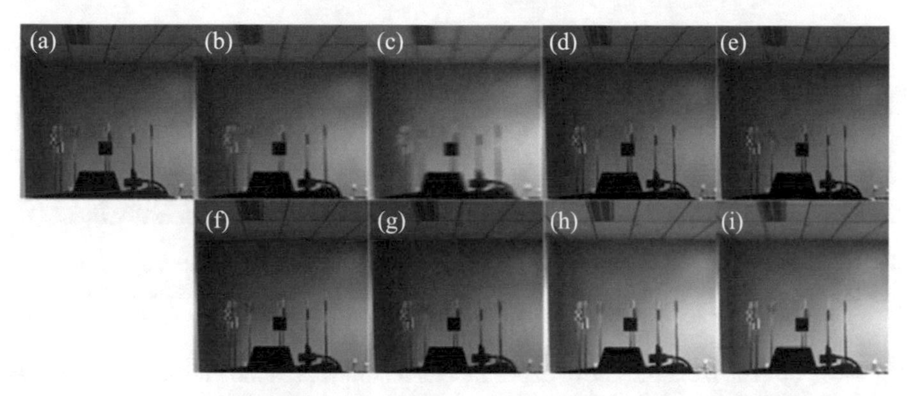

Fig. 6.16 Comparison of different restoration results **a** original image **b** blurred image with len = 5 angle = 5° **c** blurred image with len = 10 angle = 5° **d** GSR restoration results for degraded image with len = 5 angle = 5° **e** GSR [6] restoration results for degraded image with len = 10 angle = 5° **f** Fast TV restoration result for degraded image with len = 5 angle = 5° **g** Fast TV restoration result for degraded image with len = 10 angle = 5° **h** Our method restoration result for degraded image with len = 5 angle = 5° **i** Our method restoration result for degraded image with len = 10 angle = 5°

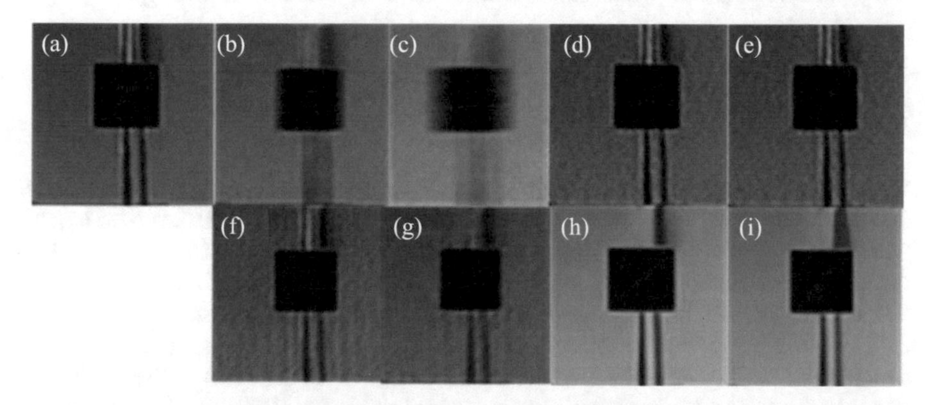

Fig. 6.17 Zoom on a 56 × 58 area extracted from the image of Fig. 6.16 **a** original image **b**) blurred image with len = 5, angle = 5° **c** blurred image with len = 5, angle = 10° **d** GSR restoration for image with len = 5, angle = 5° **e** GSR restoration for image with len = 5, angle = 10° **f** Fast TV restoration for image with len = 5, angle = 5° **g** Fast TV restoration for image with len = 5, angle = 10° **h** Our method restoration for image with len = 5, angle = 5° **i** Our method restoration for image with len = 5, angle = 10°

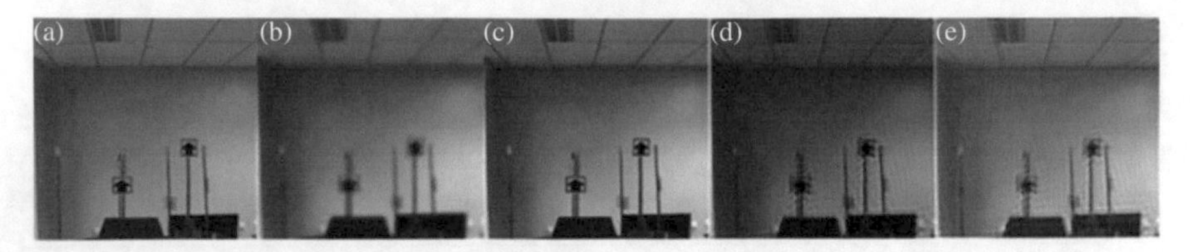

Fig. 6.18 Restoration experiment of tag1 in real application scene **a** original image **b** degraded image **c** our method restoration result **d** GSR restoration result **e** fast TV restoration result

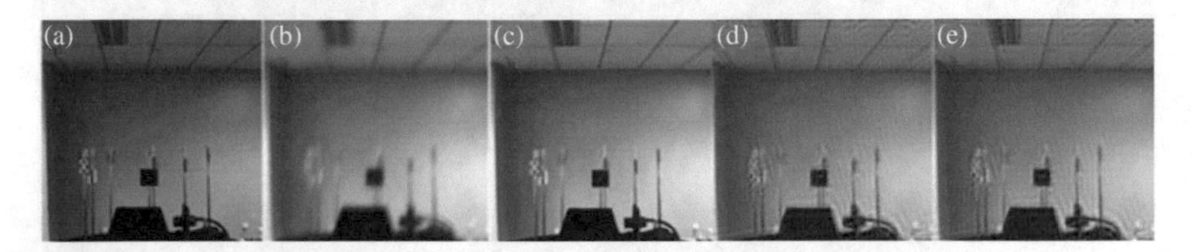

Fig. 6.19 Restoration experiment of tag2 in real application scene **a** original image **b** degraded image **c** our method restoration result **d** GSR restoration result **e** fast TV restoration result

It can be seen from the above experimental results that in the practical application scenario, the method can recover the degraded image very well. The restored images are more effective than the other two methods. The restoration results can fully meet the subsequent image matching needs.

6.5 Multi-tag 3D Position Measurement Based on Image Match

6.5.1 The Image Matching Method

In this article, the template matching method is utilized to obtain RFID tags' 3D positions. The principle of the template matching method is to look for a certain target, which has a similar size and direction to the template. By a certain algorithm, the target can be found in the large image and its coordinate position can be determined. We assume that the large image is S and the template is T. The size of S is $W \times H$. The size of T is $M \times N$ ($W > M$, $H > N$). T moves on S. The search window is denoted as S^{ij}. The coordinates of the top left corner of S^{ij} are i and j. Obviously, the search range of i and j should be $1 \le i \le H–M$, $1 \le j \le W–N$. Finally, by calculating the similarities of T and S^{ij}, we can obtain the results of template matching. The principle of template matching method is shown in Fig. 6.20.

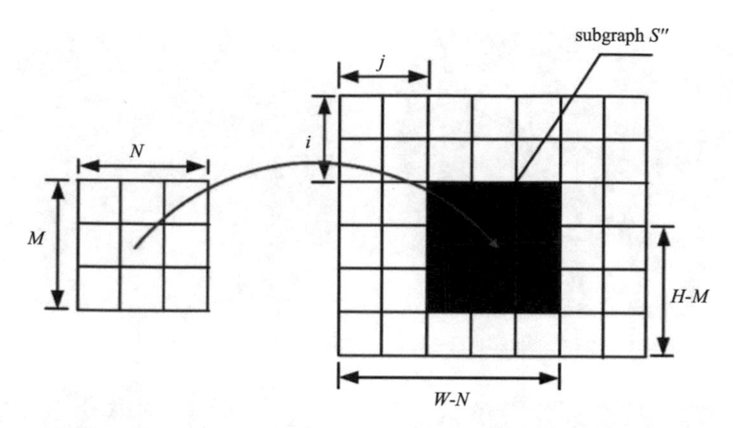

Fig. 6.20 Principle of the template matching method

6.5.2 RFID Tag 2D Position Measurement

RFID tag 2D position is measured in the position measurement module in RPMS. The vertical camera is adjusted by the control-computer to capture the images. The method of Sects. 6.4 and 6.5 is used to perform the motion blur removal processing in the images. After that, we use the iterative threshold segmentation and morphological filtering method to deal with the image to obtain the center point of a turntable. The results are shown from (a)–(c) in Fig. 6.21. Then, the horizontal camera's direction is used as the polar axis to establish the polar coordinate system, which is shown in Fig. 6.21d.

Marking points are affixed to the bottom of the tag holder. One of the markers is selected as the template. The matching process is implemented. The results of matching are shown in Fig. 6.22.

The distance from each marker to the origin is calculated. The distance from the i_{th} mark point to the origin is recorded as r_i. The angle at which the i_{th} tag rotates is θ_i. Then, the i_{th} tag's 2D position parameters are r_i and θ_i. The i_{th} tag's 2D position can be further calculated as $(r_i \cos\theta_i, r_i \sin\theta_i)$. The 2D position of every tag can be acquired by repeating the above steps.

6.5.3 RFID Tag Vertical Position Measurement

RFID tag vertical position is measured in the position measurement module in RPMS. While the turntable is rotating, one of the tags is selected as the template. The distance from the template center to the turntable in the vertical direction is measured and denoted as *h*. The template center point is selected as relative zero, and the template is matched with other tag's images to identify the positions of other tags. The matching results are shown in Fig. 6.23.

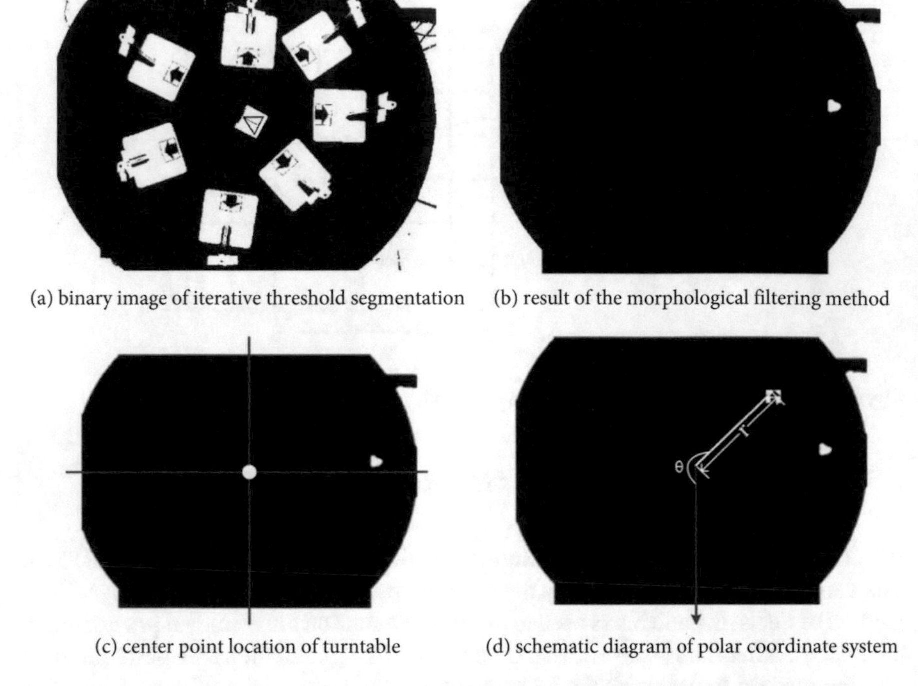

Fig. 6.21 Point location of turntable and establishment of the polar coordinate system

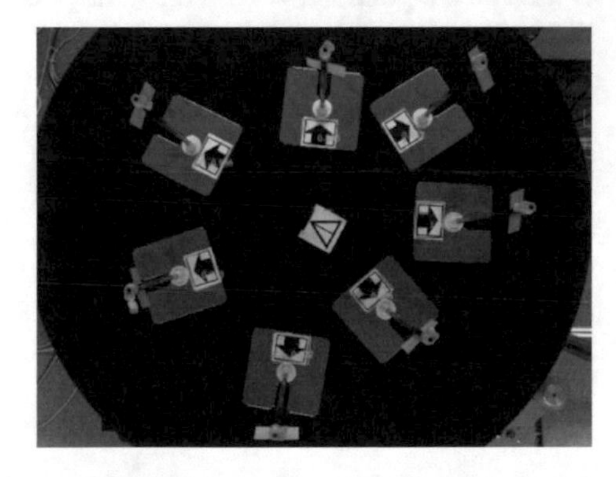

Fig. 6.22 The matching results of marking points

By the similar triangle principle, each pixel's size in the acquired image can be calculated according to the size of CCD as well as the distance between the horizontal camera and the image plane. Besides that, according to the pixel difference between the template center point and the measured tag's center point in the vertical direction, the vertical relative distance between the template and the measured tag can be calculated. Finally, the measured tag's vertical position can be calculated. The equation is shown in Eq. (6.23).

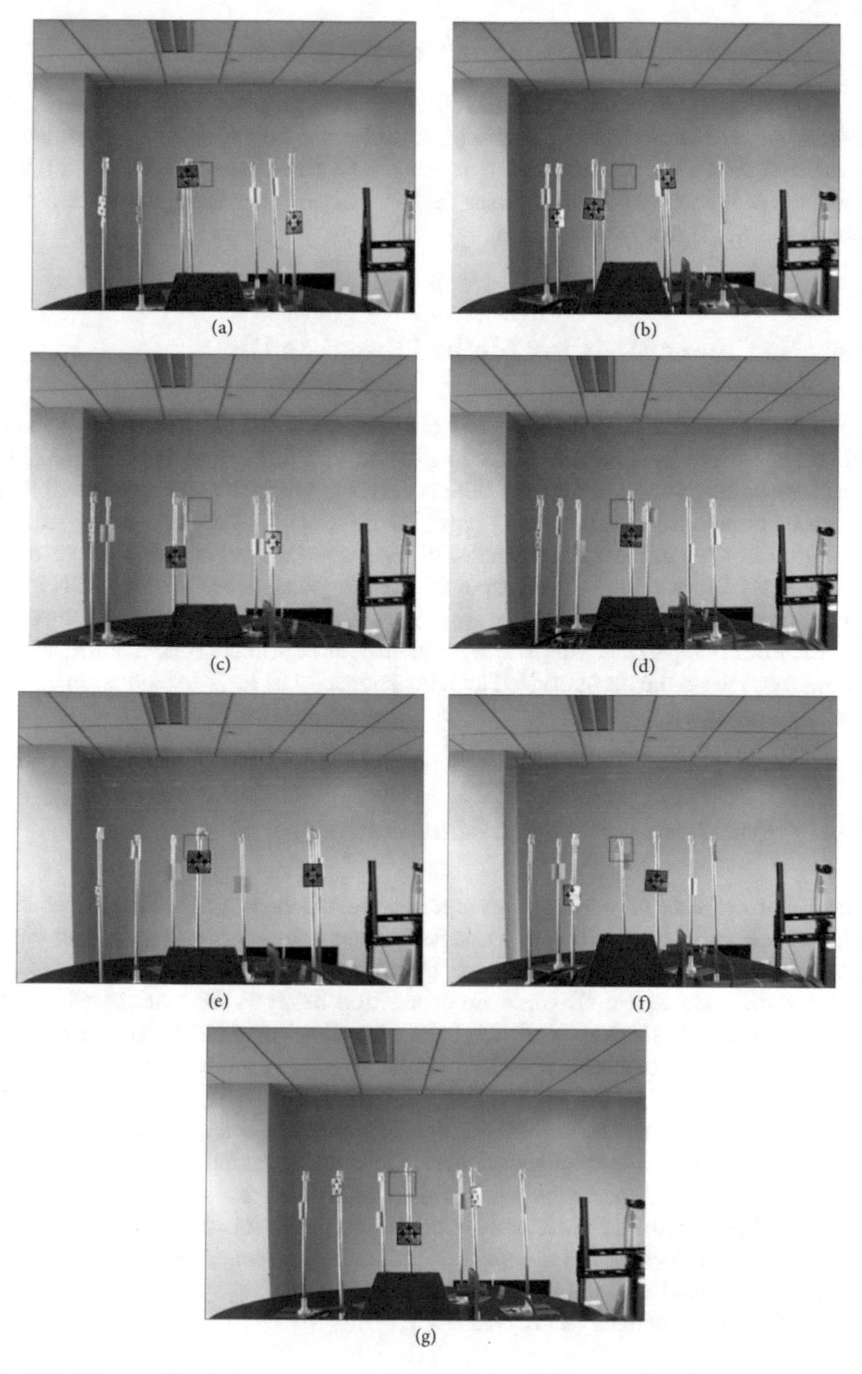

Fig. 6.23 The matching results of seven tags **a** the first tag **b** the second tag **c** the third tag **d** the fourth tag **e** the fifth tag **f** the sixth tag **g** the seventh tag

$$H_i = h \pm c_i \times a \tag{6.23}$$

In Eq. (6.23), H_i represents the i_{th} tag's vertical position. h represents the template's vertical position. c_i is the pixel difference between the template center point, and the i_{th} tag center point. a is the pixel's size in the image. When the i_{th} tag is above the template, Eq. (8) uses the " + " sign. When the i_{th} tag is below the template, Eq. (8) takes the "–" sign.

6.6 Nonlinear Modeling Method Based on DBN

According to the reading distance test module in RRRS, the reading distance of RFID tag groups is obtained. According to the first three modules in RPMS, 3D positions of RFID tag groups are obtained. By analyzing the above-obtained data, it can be concluded that the RFID tags' 3D positions and corresponding reading distance have a nonlinear relationship. In this paper, the relationship between the RFID tags' 3D positions and corresponding reading distance is built by DBN in the modeling and prediction module in RPMS. A classical DBN contains an unsupervised learning subpart and a logistic regression layer. Restricted Boltzmann machines (RBMs) construct the unsupervised learning subpart. The logistic regression layer is used for prediction.

6.6.1 Restricted Boltzmann Machine

The RBM consists of a visible layer and a hidden layer [34]. Visible variables and hidden variables are binary variables whose statuses are 0 or 1. The entire network is a bipartite graph. There are bidirectional and symmetrical connections between different layers. There is no connection between the neurons within the same layer. The RBM has excellent feature learning capabilities. The learning of features is a more abstract expression of data, which is conducive to the extraction and classification of data.

In Fig. 6.24, v_m and h_n are the node values in the visible layer and the hidden layer. W is the weight matrix that connects the visual layer and the hidden layer. b_m and c_n are the bias of the visible layer and the hidden layer.

The RBM network is described by using an energy function [32]. The energy function is used to measure the state of the entire system. The smaller the value of the energy functions, the more stable the corresponding system is. The RBM's energy function is defined as Eq. (6.24).

$$E(v, h) = -\sum_{i=1}^{n}\sum_{j=1}^{m} w_{ij}h_i v_j - \sum_{j=1}^{m} b_j v_j - \sum_{i=1}^{n} c_i h_i \tag{6.24}$$

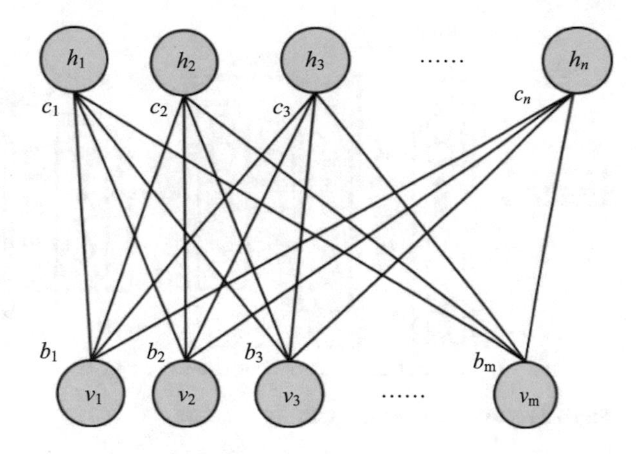

Fig. 6.24 Diagram of RBM structure

By constructing an effective energy function, the minimum value of the energy function can be determined. According to the energy function, the probabilities of visible nodes and hidden nodes can be defined as

$$p(v|h) = \frac{e^{-E(v,h)}}{\sum_h e^{-E(v,h)}} \quad p(h|v) = \frac{e^{-E(v,h)}}{\sum_v e^{-E(v,h)}} \tag{6.25}$$

By setting the parameters W, b, and c to be θ, the likelihood estimation of the joint probability of v and h can be obtained as in Eq. (6.26).

$$P_\theta(v, h) = \frac{1}{Z(\theta)} e^{-E(v,h|\theta)} \tag{6.26}$$

In Eq. (6.26), $Z(\theta)$ is the normalization factor, which is also referred to as the partition function. Combined with Eq. (6.27), the above equation can be written as

$$P_\theta(v, h) = \frac{1}{Z(\theta)} \exp(-\sum_{i=1}^{n}\sum_{j=1}^{m} \omega_{ij} h_i v_j - \sum_{j=1}^{m} b_j v_j - \sum_{i=1}^{n} c_i h_i) \tag{6.27}$$

The likelihood function $p_\theta(v, h)$ of the observation data is maximized to acquire parameters of RBMs. However, when the sample is large, traversing all possible values of v and h will make the computational complexity high. To reduce computational complexity, the input of RBMs is fitted by Gibbs sampling. Contrastive divergence (CD) is a simplified algorithm for Gibbs sampling. The advantage of the CD is that the initial training sample requires only a small number of sampling steps.

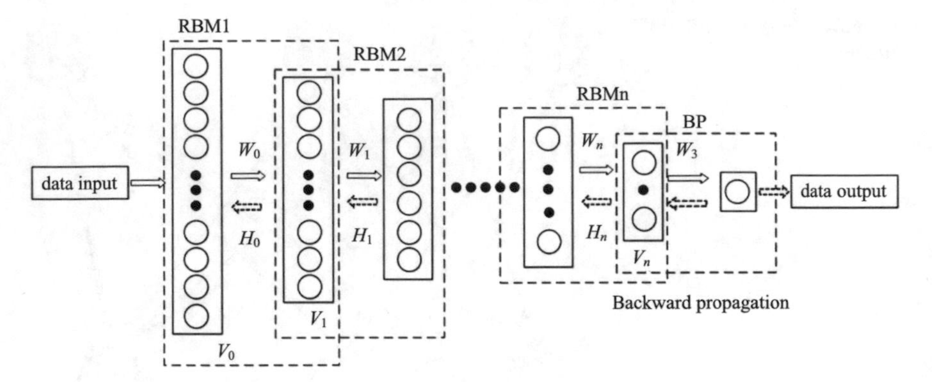

Fig. 6.25 DBN structure diagram

6.6.2 DBN

In this section, DBN consists of multi-level RBM stacks and backpropagation (BP) neural network [33, 35]. DBN is trained by layers. First, RBMs' first layer is trained with the original input data, which is expressed as v_0. By the first layer of RBMs, the v_0 is reconstructed as h_0. The reconstructed feature should preserve original feature information as much as possible. Second, the extracted feature h_0 is assigned to v_1, which is trained as the input of the second layer of RBMs. The v_1 is reconstructed as $h1$. Third, the above steps are repeated and we get v_n and h_n. The number of layers required by the network is repeatedly trained. The output of each layer of RBMs is a feature reselection process. Finally, behind the last layer of RBM, BP neural network is arranged and the connection weights are initialized. The output feature variables of the last layer of RBMs are taken as input feature variables of the BP neural network. The DBN can overcome the disadvantage of the neural network. In addition, because RBM can be quickly trained by using the CD algorithm, this framework bypasses the high complexity of directly training DBN as a whole. After training, the network is fine-tuned through the traditional backpropagation algorithm, so that the model converges optimally. The diagram of the DBN structure is shown in Fig. 6.25.

6.6.3 RFID Tag Group Model Based on DBN

In this chapter, seven tags are utilized as a set to study the influence of the spatial distribution of RFID tag groups on the corresponding reading distance. Because the input variables are 3D positions of seven tags, the number of input variables for DBN is 21. A four-layer network is used to approach the relation model. The transfer function between the hidden and visual layers in RBMs is selected as the sigmoid function. By the sigmoid function, the continuous real number can be effectively transformed into a two-value variable. The CD algorithm and gradient correction algorithm are utilized to modify the weights of the network [36, 37]. The structure of the final network is determined as [38, 39].

In this paper, seven tags are taken as a set for RFID tags' 3D position measurement. First, RRRS is utilized to acquire the RFID tag group's reading distance. Then, seven RFID tags' 3D positions at different places are acquired by the position measurement module in RPMS. The results are shown in Table 6.5. Finally, the relationship between RFID tags' 3D positions and corresponding reading distance is built by the DBN.

8950 sets of data are collected during the experiment. Among them, 8850 sets are used as the training set to train the DBN. The remaining 100 sets of data are utilized as the test set to test the DBN. The MAPE, RMSE, and PRE are utilized to evaluate the prediction performance of DBN. The definition of MAPE, RMSE, and PRE are as follows:

$$\mathrm{MAPE} = \frac{1}{N}\sum_{i=1}^{N}\left|\frac{d_p - d_r}{d_r}\right| \tag{6.28}$$

$$\mathrm{RMSE} = \sqrt{\frac{1}{N}\sum_{i=1}^{N}(d_p - d_r)^2} \tag{6.29}$$

$$\mathrm{PRE} = \frac{|d_p - d_r|}{|d_r|} \tag{6.30}$$

In Eq. (6.28), (6.29), and (6.30), d_p is the forecasted value and d_r is the actual value.

The reading distance of the remaining 100 sets of data predicted by DBN is shown in Fig. 6.26.

To evaluate the DBN, the prediction results of DBN are compared with GA-BP and PSO-BP. The comparison results are as follows:

As we can see from Fig. 12 and Table 6.6, the relationship between RFID tags' 3D positions and reading distance can be well established by DBN. The MAPE and

Table 6.5 RFID tag group's 3D positions and corresponding reading distance

Sample ID	x_1/m	y_1/m	z_1/m	…	x_7/m	y_7/m	z_7/m	d_r/m
1	0.815	0.278	0.216	…	0.008	0.221	0.314	1.230
2	0.906	0.547	0.189	… …	0.048	0.184	0.314	1.790
3	0.127	0.957	0.117	…	−0.189	0.093	0.284	0.960
⋮	⋮	⋮	⋮		⋮	⋮	⋮	⋮
8948	0.712	0.173	0.300	…	−0.003	0.053	0.264	1.200
8949	−0.210	0.209	0.433	…	−0.075	0.166	0.311	1.800
8950	−0.195	0.796	0.507	…	0.069	0.121	0.329	1.810

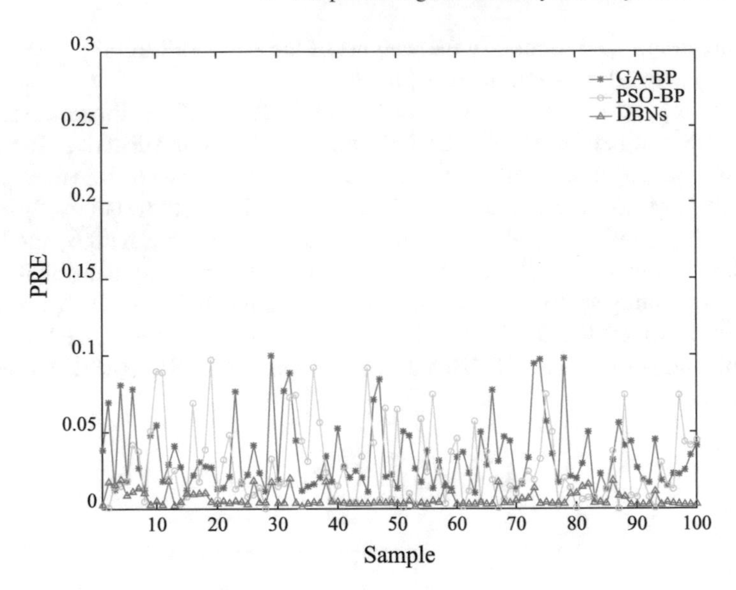

Fig. 6.26 Prediction results of different methods

Table 6.6 Prediction evaluation results of different methods

Methods	MAPE/%	RMSE
GA-BP	3.440	0.398
PSO-BP	2.770	0.290
DBN	1.640	0.168

RMSE of DBN are 1.64% and 0.168 respectively. Compared with GA-BP and PSO-BP, the MAPE and RMSE of DBN are smaller and the prediction performance is better.

Besides that, it can also be seen from the above results that the DBN can forecast the tag groups' reading distance and find out the optimal distribution structure of tag groups corresponding to the maximum reading distance effectively. The maximum prediction reading distance in the above remaining 100 sets of data is 2.3671, and its corresponding optimal distribution structure of the tag group is shown in Fig. 6.27.

6.7 Conclusion

RFID multi-tag 3D measurement system based on deep learning is designed in this chapter. During the procedure of measurement, in order to improve the measurement accuracy, the noise in the obtained images are removed by flexible denoising convolutional neural network (FDnCNN), and the motion blurs and noise in the obtained images are removed by using the Wiener filtering and MWCNN. After

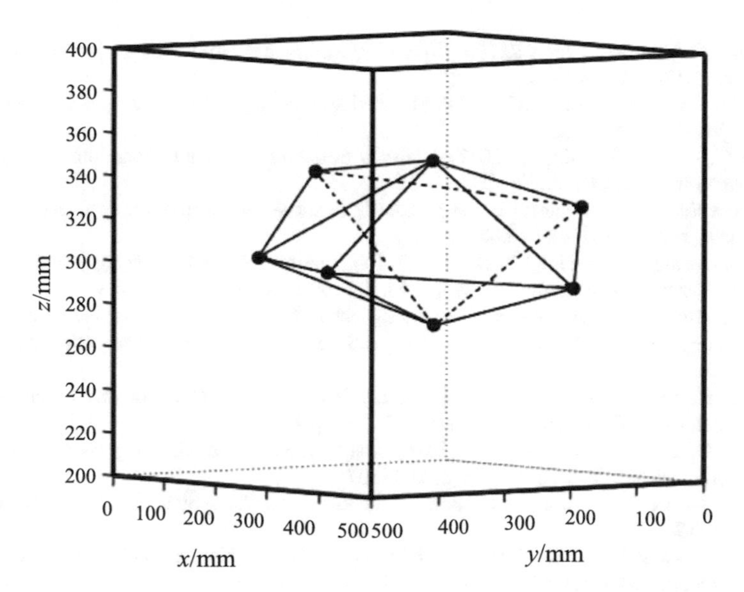

Fig. 6.27 Optimal structure of RFID tag network

the image is restored, the image matching method is used to obtain the 3D coordinates of the tag group. Due to the nonlinear relationship between the 3D coordinates of RFID tags and the corresponding reading distance, we use DBN to model the nonlinear relationship. The established model is applied to forecast the RFID tag group's reading distance. The experimental results show that the MAPE and RMSE of DBN are 1.64% and 0.168 respectively. Compared with GA-BP and PSO-BP, the MAPE of DBN can reduce about 40.8% and 52.3%, respectively, and the RMSE of DBN can reduce about 42% and 57.8%, respectively. The MAPE and RMSE of DBN are smaller. The DBN can better model the nonlinear relationship between the RFID tags' 3D positions and reading distance. The established DBN can predict the reading distance of unknown tag groups and find out the optimal distribution structure of tag groups corresponding to the maximum reading distance. The proposed method can provide guidance for the geometric distribution of tag groups in practical application scenarios.

References

1. Martinelli F (2015) A Robot Localization System Combining RSSI and Phase Shift in UHF-RFID Signals. IEEE T Control Syst T 23(5):1782–1796
2. Liu X, Yang Q, Luo J et al (2019) An Energy-aware Offloading Framework for Edge-augmented Mobile RFID Systems. IEEE Internet Things 6(3):3994–4004
3. Gope P, Amin R, Hafizul SK et al (2018) Lightweight and privacy-preserving RFID authentication scheme for distributed IoT infrastructure with secure localization services for smart city environment. Future Gener Comput Syst 83:629–637

4. Hinton GE, Rsalakhutdinov R (2006) Reducing the dimensionality of data with neural networks. Science 313(5786):504–507
5. Hinton GE, Osindero S, Teh YW (2006) A fast learning algorithm for deep belief nets. Neural Comput 18(7):1527–1554
6. Liu W, Wang Z, Liu X et al (2017) A survey of deep neural network architectures and their applications. Neurocomputing 234:11–26
7. Litjens G, Kooi T, Bejnordi BE et al (2017) A survey on deep learning in medical image analysis. Med Image Anal 42:60–88
8. Faust O, Hagiwara Y, Hong TJ et al (2018) Deep learning for healthcare applications based on physiological signals: a review. Comput Meth Programs Biomed 161:1–13
9. Lefkimmiatis S (2018) Universal denoising networks: a novel CNN architecture for image denoising. In: IEEE Conference on Computer Vision and Pattern Recognition (CVPR), pp 3204–3213.
10. Park JH, Kim JH, Cho SI (2018) The analysis of CNN structure for image denoising. In: International SoC Design Conference (ISOCC), pp 220–221
11. Dong C, Loy CC, He KM et al (2016) Image super-resolution using deep convolutional networks. IEEE T. Pattern Anal 38(2):295–307
12. Davy A, Ehret T, Facciolo G et al, Non-Local Video Denoising by CNN. https://arxiv.org/abs/1811.12758
13. Yu XL, Wang DH, Zhao ZM (2019) Semi-physical verification technology for dynamic performance of internet of things system. Springer Singapore.
14. Yu YS, Yu XL, Zhao ZM et al (2018) Image analysis system for optimal geometric distribution of RFID tags based on flood fill and DLT. IEEE T Instrum Meas 27(4):839–848
15. Tassano M, Delon J, Veit T, An Analysis and Implementation of the FFDNet Image Denoising Method, Image Processing On Line (IPOL)
16. Zhang K, Zuo WM, Zhang L (2018) FFDNet: Toward a Fast and Flexible Solution for CNN-Based Image Denoising. IEEE T Image Process 27(9):4608–4622
17. Gshuhang U et al (2014) Weighted nuclear norm minimization with application to image denoising. In: Proceedings of the IEEE conference on computer vision and pattern recognition, pp 2862–2869
18. Zhang K, Zuo WM, Gu SH et al (2017) Learning deep CNN denoiser prior for image restoration. In: IEEE Conference on Computer Vision and Pattern Recognition (CVPR), pp 3929–3938
19. Zhang K, Zuo WM, Chen YJ et al (2017) Beyond a gaussian denoiser: residual learning of deep CNN for image denoising. IEEE T Image Process 26(7):3142–3155
20. Dash R, Majhi B (2014) Motion blur parameters estimation for image restoration. Optik - Int J Light Electron Opt 125(5):1634–1640
21. Wang Z, Yao Z, Wang Q (2017) Improved scheme of estimating motion blur parameters for image restoration. Digit Signal Prog 65:11–18
22. Takagi Y, Fujisawa T, Ikehara M (2017) Image restoration of JPEG encoded images via block matching and wiener filtering, IEICE Trans Fundam Electron Commun Comput Sci 100(9):1993–2000
23. Brifman A, Romano Y, Elad M (2016) Turning a denoiser into a super-resolver using plug and play priors. In: 2016 IEEE International Conference on Image Processing (ICIP). IEEE, pp 1404–1408
24. Chan SH, Wang X, Elgendy OA (2016) Plug-and-play ADMM for image restoration: fixed-point convergence and applications. IEEE Trans Comput Imaging 3(1):84–98
25. Romano Y, Elad M, Milanfar P (2017) The little engine that could: Regularization by denoising (RED). SIAM J Imaging Sci 10(4):1804–1844
26. Zoran D, Weiss Y (2011) From learning models of natural image patches to whole image restoration. In: 2011 international conference on computer vision. IEEE, pp 479–486
27. Heide F, Steinberger M, Tsai YT et al (2014) FlexISP: A flexible camera image processing framework. ACM Trans Graph (TOG) 33(6):231
28. Chambolle A, Pock T (2011) A first-order primal-dual algorithm for convex problems with applications to imaging. J Math Imaging Vis 40(1):120–145

29. Kingma DP, Ba J Adam (2014) A method for stochastic optimization. https://arxiv.org/abs/1412.6980
30. Zhang J, Zhao D, Gao W (2014) Group-based sparse representation for image restoration. IEEE Trans Image Process 23(8):3336–3351
31. Ren D, Zhang H, Zhang D et al (2015) Fast total-variation based image restoration based on derivative alternated direction optimization methods. Neurocomputing 170:201–212
32. Abdel-Zaher AM, Eldeib AM (2016) Breast cancer classification using deep belief networks. Exp Syst Appl 46:139–144
33. Fu G (2018) Deep belief network based ensemble approach for cooling load forecasting of air-conditioning system. Energy 148:269–282
34. Kuremoto T, Kimura S, Kobayashi K et al (2014) Time series forecasting using a deep belief network with restricted Boltzmann machines. Neurocomputing 137:47–56
35. Chong E, Han C, Park FC (2017) Deep learning networks for stock market analysis and prediction, methodology, data representations, and case studies. Exp Syst Appl 83:187–205
36. Gaukler G, Ketzenberg M, Salin V (2017) Establishing dynamic expiration dates for perishables: An application of rfid and sensor technology. Int J Prod Econ 193:617–632
37. Chen JC, Cheng CH, Huang PTB et al (2013) Warehouse management with lean and RFID application: a case study. Int J Adv Manuf Technol 69(1–4):531–542
38. Yu Y, Yu X, Zhao Z et al (2016) Measurement uncertainty limit analysis of biased estimators in RFID multiple tags system. IET Sci Meas Technol 10(5):449–455
39. Yadav AK, Schandel S (2014) Solar radiation prediction using Artificial Neural Network techniques: a review. Renew Sustain Energy Rev 33:772–781

Printed in the United States
by Baker & Taylor Publisher Services